CLIMATE CHANGE IN THE GREAT LAKES REGION

CLIMATE CHANGE IN THE GREAT LAKES REGION

Navigating an Uncertain Future

Edited by Thomas Dietz and David Bidwell

Michigan State University Press

East Lansing

☉ The paper used in this publication meets the minimum requirements of ANSI/NISO
Z39.48-1992 (R 1997) (Permanence of Paper).

Michigan State University Press
East Lansing, Michigan 48823-5245

Printed and bound in the United States of America.

18 17 16 15 14 13 12 1 2 3 4 5 6 7 8 9 10

LIBRARY OF CONGRESS CATALOGING-IN-PUBLICATION DATA

Climate change in the Great Lakes region : navigating an uncertain future / edited by
Thomas Dietz and David Bidwell.
    p. cm.
    Includes bibliographical references.
    ISBN 978-1-61186-012-2 (cloth : alk. paper) 1. Climatic changes—Great Lakes Region
(North America) 2. Great Lakes Region (North America)—Environmental conditions. I.
Dietz, Thomas. II. Bidwell, David, 1969–

    QC902.2.G74C57 2011
    551.6977—dc22   2011008830

Cover design by Erin Kirk New
Book design by Scribe Inc. (www.scribenet.com)

g green press INITIATIVE Michigan State University Press is a member of the Green Press Initiative and
is committed to developing and encouraging ecologically responsible publishing
practices. For more information about the Green Press Initiative and the use of recycled
paper in book publishing, please visit www.greenpressinitiative.org.

Visit Michigan State University Press on the World Wide Web at:
www.msupress.msu.edu

Dedicated to Stephen H. Schneider
February 11, 1945–July 19, 2010

Contents

Thinking about Climate Change in the Great Lakes Region   1

PART ONE. EFFECTS OF CLIMATE CHANGE
IN THE GREAT LAKES REGION

Historical Climate Trends in Michigan and the Great Lakes Region
    Jeffrey A. Andresen   17

Climate Change Impacts and Adaptation Strategies for Great Lakes
    Nearshore and Coastal Systems
    Scudder D. Mackey   35

Climate Change and Biodiversity in the Great Lakes Region: From
    "Fingerprints" of Change to Helping Safeguard Species
    Kimberly R. Hall and Terry L. Root   63

PART TWO. DECISION MAKING AND CLIMATE CHANGE

Decision Making under Climate Uncertainty: The Power of
    Understanding Judgment and Decision Processes
    Sabine M. Marx and Elke U. Weber   99

Agricultural Adaptation to Climate Change: Is Uncertain Information
    Usable Knowledge?
    William E. Easterling, Clark Seipt, Adam Terando, and
    Xianzeng Niu   129

Adapting to Climate Change in the Context of Multiple Risks: A Case
    Study of Cash Crop Farming in Ontario
    Ben Bradshaw, Suzanne Belliveau, and Barry Smit   159

PART THREE. ADAPTATION TOOLS AND CASE STUDIES

The Contextual Importance of Uncertainty in Climate-Sensitive
   Decision-Making: Toward an Integrative Decision-Centered
   Screening Tool
         Susanne Moser   179

Linking Science to Decision Making in the Great Lakes Region
         Joel D. Scheraga   213

The Development and Communication of an Ensemble of Local-Scale
   Climate Scenarios: An Example from the Pileus Project
         Julie A. Winkler, Jeanne M. Bisanz, Galina S. Guentchev,
         Krerk Piromsopa, Jenni van Ravensway, Haryono
         Prawiranata, Ryan S. Torre, Hai Kyung Min, and
         Johnathan Clark   231

Preparing for Climate Change in the Great Lakes Region   249

Notes on the Contributors   259

Reflections on Stephen H. Schneider   265

Index   267

Thinking about Climate Change in the Great Lakes Region

THINKING ABOUT CLIMATE CHANGE MADE A DRAMATIC TURN IN THE first decade of the twenty-first century. The scientific consensus is a near certainty that human activities are changing the climate—that warming is unequivocal. To quote a recent report by the U.S. National Academies of (2010b, 1): "A strong, credible body of scientific evidence shows that climate change is occurring, is caused largely by human activities, and poses significant risks for a broad range of human and natural systems." The full report (U.S. National Research Council 2010b) goes on to call for a "new era of climate change research," one in which science contributes to fundamental knowledge but also supports the myriad decisions being made in response to climate change by local, state/provincial, tribal, and national governments; small businesses and large corporations; and households and individuals.

Researchers from a variety of disciplines are responding to the new era with work focused on what a changing climate means to the health and well-being of the planet. There is a growing awareness that it poses unprecedented dangers to ecologies, cultures, and economies on a global scale. Popular media warns of potentially dire consequences of climate change: desertification of previously fertile land, outbreaks of disease, powerful tropical storms, loss of glaciers and polar icecaps, and flooding of coastal areas. Political debate centers on how best to slow or stop the emission of gases that contribute to climate change, and who should be responsible for these mitigative actions.

Unfortunately, two important points are only recently being incorporated into the research enterprise. First, even if we significantly curtail the burning of fossil fuels, greenhouse gases previously pumped into the atmosphere will continue to affect the climate for centuries to come (Solomon et al. 2009,

U.S. National Research Council 2010c). Moreover, the rise in global energy demand is outstripping development of alternative energy sources and efficiency increases, meaning that coal, gas, and oil, with attendant greenhouse gas emissions, will continue to serve as major energy sources into the foreseeable future (U.S National Research Council 2010e). Models indicate that under even the most optimistic scenarios, the world's climate will continue to warm for the next several decades. Recent analyses suggest that it is unlikely that any policies adopted in the near term will hold the planet to the 3.6°F (2°C) of average warming that has been identified as the guardrail to prevent "dangerous interference" with the climate (Fawcett et al. 2009). So, substantially greater risk from climate change seems inevitable.

Second, we are already feeling the effects of climate change (U.S. National Research Council 2010b). Scientists have documented significant changes in local weather patterns. Heat waves have become longer and more extreme, while cold snaps have become shorter and milder. Snow cover is decreasing, and rivers and lakes are freezing later and thawing earlier. The amounts and timing of rain and snowfall are more variable. As a result of these changes, the ranges of plants and animals are shifting. Admittedly less dramatic than popular doomsday images, these changes have real and serious consequences for the ecological systems that support life—and human economies. While it is critical to address the causes of climate change, we must also acknowledge that change is occurring and is likely to continue.

Although early work on climate change focused primarily on how to limit human influences on climate, more recent work has taken on the importance of *adaptation* as a response to climate changes (Grothmann and Patt 2005; Smit et al. 2000; Kasperson, Kasperson, and Turner 2010). "Adaptation is generally perceived to include an adjustment in social-ecological systems in response to actual, perceived, or expected environmental changes and their impacts" (Janssen and Ostrom 2006, 237). In the context of climate change, adaptation refers to "a process, action or outcome in a system (household, community, group, sector, region, country) in order for the system to better cope with, manage or adjust to some changing condition, stress, hazard, risk or opportunity" (Smit and Wandel 2006). The production of national reports for Canada (Natural Resources Canada 2004), the United Kingdom (Willows and Connell 2003), Australia (Allen Consulting Group 2005) and the United States (Karl, Melilo, and Peterson 2009; U.S. National Research Council 2010a) illustrates the growing awareness of the need to address adaptation to climate change. Indeed, it is now widely recognized that adaptation must be a major element in U.S. response to climate change, and that effective adaptation must be grounded in research of the new era of climate change—research based in the best science, and responsive to the needs of decision makers (U.S. National Research Council 2010a).

We regularly adapt to changes in the weather. When the forecast calls for rain, we can choose to carry an umbrella, put on a raincoat, stay indoors, or get wet. When temperatures dip, we can put on a sweater, light a fire, or shiver. But how do we adapt to the kinds of changes that might come from climate change? For example, how does a city planner prepare for more days of heavy rain? When should a farmer plant crops if the date of the last frost becomes less predictable? How much should the owner of a ski lodge invest in new equipment if snowfall is less reliable? How does a conservation agency preserve critical habitat for an endangered songbird if forest types are shifting their boundaries?

To protect ourselves and the things we value against current and future effects of climate change, we will have to answer these questions. Unfortunately, no single answer will serve as a silver bullet. Adaptation to climate change is complicated by significant uncertainty and complexity in coupled human and natural systems (Liu et al. 2007a, 2007b; Rosa et al. 2010). While science is quite certain that the climate is changing, researchers still have much to learn about how local and regional conditions will be affected by this change. Weather is a complex system, influenced by subtle differences in geography. Moreover, how humans balance growing energy needs and efforts to mitigate the human role in climate change will influence the magnitude of the changes we face. Coupled human and natural systems are also complex, so it is difficult to predict exactly how plants, wildlife, and people will respond to changes in temperatures and precipitation and other climate effects. Variability in the resilience and vulnerability of different species and economic sectors means that climate change will create minor annoyances in some systems, new opportunities for some, and dire crises for others. People are not always good at coping with this kind of complexity and uncertainty, which adds further challenges to successful climate change adaptation.

A unique gathering of scientists, resource managers, businesspeople, and policymakers met in March 2007 at the East Lansing campus of Michigan State University to consider these issues. With funding from the National Science Foundation and MSU's Environmental Research Initiative, the university's Environmental Science and Policy Program hosted a two-day symposium entitled *Climate Change in the Great Lakes Region: Decision Making under Uncertainty.* A group of distinguished scholars presented papers on one of three topics: (1) the anticipated effects of climate change in the Great Lakes region, (2) possible adaptations to those effects, and (3) how humans make decisions under uncertainty. Updated versions of their papers comprise the chapters of this volume.

We organized these papers into three parts reflecting the three symposium topics. However, we also acknowledge that many of the papers address several important topics and could have been included in any one of the three parts.

EFFECTS OF CLIMATE CHANGE IN THE GREAT LAKES

Where effective mitigation of climate change causes is generally discussed at a national or global scale, the effects of climate change are best addressed at smaller scales (Grothmann and Patt 2005), including the regional, local, and even individual levels. Many studies in the adaptation literature focus on the regional scale. Regions are generally defined by similar geographies, ecologies, and economic activities, as well as common social and cultural traits. This book (and the symposium from which its chapters sprang) is focused on the Great Lakes region of North America. This includes portions of several U.S. states, including Illinois, Indiana, Michigan, Minnesota, New York, Ohio, Pennsylvania, and Wisconsin, as well as the Canadian province of Ontario.

The State of Michigan, central to the region, served as a common example in discussions and the papers. In addition to his role as a member of the MSU faculty, Jeffrey Andresen serves as the state climatologist for Michigan. In his chapter, he provides a historical overview of climate trends in Michigan, focused primarily on changes over the most recent decades. Andresen presents a picture of Michigan weather that is warmer (especially in winter and early spring) and wetter than we have experienced in the past. This new weather pattern is linked to a profound reduction in wintertime freezing of the Great Lakes, which increases the amount of moisture released into the atmosphere, increasing the amount of lake effect snow.

The other two chapters in this first part discuss how climate change affects the ecology and ecological services in the region. Scudder Mackey discusses how water levels in the Great Lakes could drop due to increased evaporation and changes in precipitation. These changes could seriously endanger the integrity of nearshore habitat for fish and other wildlife. Moreover, structural and chemical changes in substrates might lead to greater colonization by invasive species, already a significant stressor on the ecology of the lakes.

Focusing primarily on one type of anticipated change—temperature increase—Kimberly Hall and Terry Root discuss how both terrestrial and aquatic plants and animals could respond to climate change. These responses include shifts in home ranges, population densities, and timing of critical behaviors (e.g., migration and breeding, called phenology). These shifts would pose substantial ecological risks, because they could lead to mismatches in age-old relationships among species. For example, if migrating birds arrive early to the region, they might not find the insects they need to refuel. All of these factors could lead to the selective survival of particular physical traits and behaviors within species—or to local extinctions. Both Mackey, and Root and Hall discuss the difficulties these changes pose to the conservation of vulnerable species and habitats. They advise that mangers seek ways to improve the resilience of these systems so they are better prepared for expected and anticipated changes.

DECISION MAKING UNDER UNCERTAINTY

The theme of the second part of this volume is making decisions under conditions of uncertainty—sometimes called *adaptive risk management* (Arvai et al. 2006; Renn 2008; U.S. National Research Council 2010b). We cannot know with certainty what the future will hold, but experience and careful research means that our understanding of the effects of climate change will change over time. As with so many other aspects of life, coping with climate change involves risk management. Acknowledging that we will learn more as we move forward suggests that our approach to managing climate risks should change over time, thus *adaptive* risk management. Our authors are particularly concerned with the challenge of integrating scientific knowledge into the decision process—the heart of adaptive risk management. Two common lessons in these chapters are that scientists need to focus on the needs of decision makers, and that there is great variability in the kinds of decisions to be made and the contexts in which decisions are made.

There is still a great deal not known about how changes in the climate will affect patterns of weather at local and regional scales. These uncertainties make it difficult to know exactly how one will need to adapt. "Uncertainties about climate change not only shape international, national and local climate policy, but they also influence perceptions of and responses to climate change at the level of individuals, communities, and businesses" (Dessai, O'Brien, and Hulme 2007, 1). Fortunately, scholars have begun to apply broad themes to how people understand and respond to uncertainty in the challenges of climate change (Grothmann and Patt 2005; Marx et al. 2007).

Among these themes is the nature of human cognition. In their chapter, Sabine Marx and Elke Weber provide an introduction to how humans respond to uncertainty. Drawing from scholarship in the decision sciences and risk perception literatures, they explain various cognitive biases that are likely to influence decision makers' responses to climate change. They assert that humans "have a great need for predictability," but uncertainty impairs the ability to predict outcomes. There are two modes of thought through which humans can cope with the anxiety that comes from uncertainty. One mode is analytic and deliberate, in which people weigh evidence and predict probabilities. The more prevalent mode is experiential and affective. This experiential mode of thought operates by a series of rules of thumb, or heuristics. Marx and Weber also elucidate how climate-related decisions are affected by people's limited abilities to cope with choices with delayed outcomes. They synthesize these threads of evidence with discussion of how climate risks could be most effectively communicated, and how climate decisions could be framed, to reduce the chances of the kinds of errors in decision making they have reviewed.

It is not just human nature that complicates decision making. The scientific models we use to understand the effects of climate change are inevitably uncertain. William Easterling, Clark Seipt, Adam Terando, and Xianjeng Niu provide an overview of the sources of uncertainty in models used to project climate impacts. They use the problem of how climate change could threaten global food security as their test bed. This is a particularly salient concern for the Great Lakes region, given the prominent role of agriculture in the economy of the area. They describe three kinds of uncertainty that challenge our understanding of climate change impacts: fundamental, structural, and parametric. These uncertainties pose challenges to modeling the appropriate responses of agriculture to climate change. Fundamental uncertainty occurs when the nature or novelty of an issue makes modeling intractable. Ambiguous or incomplete models result in structural uncertainty as source of disagreement among different climate models. Parametric (value) uncertainty is caused by missing, inaccurate, or otherwise inadequate data, which is likely to affect climate change models due to the many forms of data they require. Easterling et al. propose that successful adaptation will require a stream of "usable knowledge," which they define as information that is tractable to users, credible, legitimate, and salient. Achieving this will require a close relationship between farmers and the adaptation research community.

Ben Bradshaw, Suzanne Belliveau, and Barry Smit take a different approach to understanding how stakeholders make decisions under uncertainty. Specifically, they assess ways in which Ontario farmers currently confront climate-related and other kinds of uncertainty. For example, farmers already must adapt to years with too little or too much rain. Bradshaw et al. found that these farmers rely on crop insurance as a safety net for the most dire situations, but also employ both proactive and reactive management strategies (e.g., exchanging seed varieties when a wet spring reduces the growing time for their crops). Their interviews with farmers also remind us that stakeholders cope with uncertainties not related to climate. In the case of farmers, these uncertainties include commodity prices and market demand. These other factors might contribute to or diminish the adaptive capacity of a decision maker.

MODELS FOR ADAPTIVE DECISION MAKING

Our final three chapters provide examples of successful approaches of linking uncertain science with decision making (see also U.S. National Research Council 2010d). There is a tremendous range of problems being faced by decision makers. In her chapter on integrating science into decisions regarding climate, Susanne Moser stresses the need for scientists to understand the specific needs of decision makers. Moser provides a three-dimensional typology of decisions. One axis differentiates between goal-oriented decisions and evaluative decisions. The second axis is the time horizon of the decision—short-term versus long-term. The third dimension is the finality of the decision—is it a one-time choice, or can the choice be revisited? Moser contends that the effects of uncertainty depend, in part, on the type of decision being faced. She offers one potential tool—the Decision Uncertainty Screening Tool, or DUST—for determining the kinds of uncertainty analyses required for a given decision.

Joel Scheraga summarizes two efforts undertaken by the U.S. Environmental Protection Agency's Global Change Research Program. Acknowledging the local scale of adaptation, these initiatives provide information that stakeholders can use in conducting place-based assessments of vulnerability to climate change effects. Given the uncertainty of current climate models, some researchers have suggested that local assessments are worthless. Scheraga demonstrates how communities can tackle important questions regarding the sensitivity and vulnerability of local infrastructure (e.g., water and sewage management resources) to a range of potential effects. These projects began by engaging stakeholders in a "decision inventory," which helped to determine priorities for assessment and action given limited resources for climate adaptation.

The Pileus Project at Michigan State University, introduced by Julie Winkler and colleagues, created a web-based climate projections tool for the agriculture and tourism industries in Michigan. By downscaling data from climate models, the project was able to create scenarios that communicate uncertainty at a scale that is meaningful to decision makers. Most notable in this case is the involvement of stakeholders in determining the parameters of the scenarios and designing a number of "user cases," which allow users to explore data most relevant to their needs.

A common thread running through most of the chapters of this volume is the need to focus science on the needs of decision makers. This constitutes a mode of science that Smit and Wandel (2006, 285) have called *practical application*, defined as "research that investigates the adaptive capacity and adaptive needs in a particular region or community in order to identify means of implementing adaptation initiatives or enhancing adaptive capacity." Scheraga, for example, stresses the importance of engaging stakeholders at the outset of a climate-change assessment process, in order to ensure the development of resources appropriate to needs of end users.

Vogel et al. (2007) have proposed that fostering successful adaptation to climate change requires new relationship between scientists and practitioners. They note, however, that scientists have often done a poor job communicating with policymakers and managers. In fact, there are few opportunities for face-to-face interaction between scientists and the practitioners who will use their research findings. Easterling et al. explain, "The missing catalyst for action is a vigorous discourse between scientists and farmers over the uncertainties surrounding adaptation research."

A new style of communication between scientists and practitioners has been evolving. "It is a more democratic model of communication in which different experts, risk-bearers, and local communities all have something to bring to the table" (Vogel et al. 2007, 360). Moser notes a potential role of her DUST model as a boundary object around which scientists and decision makers can interact. These kinds of interactions would serve to enhance the credibility, legitimacy, and relevance of information being developed and communicated by scientists.

In some ways, the March 2007 symposium that led to this book operated as another such boundary object. When we began planning for the event, we hoped to attract up to 100 people. We were pleasantly surprised by the active participation of a diverse group of nearly 200 attendees. Active outreach to a variety of organizations and sectors resulted in the turnout of a good mix of local and state policymakers, university scientists and students, and representatives of resource management agencies, utilities, and agricultural and business organizations. Following each of three sets of speakers' presentations, small panels of stakeholders discussed the papers, and we engaged the full group of participants in facilitated discussions. The results were rich, cross-disciplinary discussions during the formal sessions, as well as during coffee breaks and meals.

At the close of the meeting, we held a working session of our speakers and other key participants with a goal of articulating "lessons learned" at the symposium. A year later, we held an open meeting on climate change in the Great Lakes region, which attracted 160 attendees in a dozen topic-specific sessions. Here, too, we closed the meeting with a working session to identify "lessons learned." The final chapter of this volume is an extract of those lessons. It constitutes a synthesis of the views of scientists and decision makers from the public and private sectors, as well as educators and citizens.

In addition to the synthesis of lessons learned, we believe the symposium and the activities leading up to and following it reflect a model for how to conduct successful research on climate change adaptation (see figure 1).

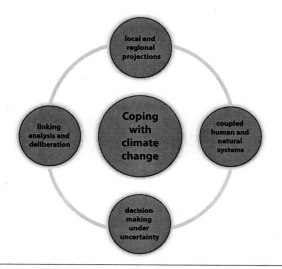

Figure 1. Successful research on climate change requires four kinds of expertise to produce effective analysis and action.

The approach, pioneered by the Pileus Project at MSU, requires four kinds of expertise to produce effective analysis and action plans: making regional and local projections of effects; considering coupled human and natural systems; addressing decision making under uncertainty; and linking technical analysis and stakeholder deliberation. Each aspect of the model is discussed in more detail below.

Bring Climate Change Home

Climate change plays out at the local level, but models of climate are global and regional and do not provide the detailed information needed for most on-the-ground decision making. There are a variety of tools available, however, that produce effective local and regional projections based on the global climate models. Thus, a critical step in research to support decision making is to use "downscaling" to provide projections of key climate/weather variables at the local and regional levels.

Engage Expertise on Coupled Human and Natural Systems

Another critical step is to engage the best available expertise on systems that will be impacted. Climate change will impact many coupled human and natural systems and key economic sectors such as agriculture, tourism, and civic infrastructure. Researchers and stakeholders from a variety of fields already have substantial expertise in those systems, which we can engage in understanding how climate will interact with the other stressors on those systems and change their dynamics. This will help inform the downscaling efforts by identifying what climate factors may matter most, while the local and regional forecasts provide information critical to understanding the climate regimes' key systems.

Deploy Expertise on Decision Making under Uncertainty

A high degree of uncertainty about both climate trajectories and the dynamics of coupled human and natural systems is inevitable. Individuals, groups, communities, organizations, and governments have to make decisions in the face of this uncertainty. Decision sciences, economics, political science, psychology, and sociology all have considerable understanding of decision making and have developed tool kits to aid in effective decision making. Expertise in these areas has to inform the full suite of climate change activities.

Link Scientific Analysis with Deliberation

Deliberation with interested and impacted parties—decision makers—is essential from the start. Such engagement is essential for "getting the right science and getting the science right." It is also essential for building understanding of, and trust in, the science. Such interaction enhances societal capacity for decision making under uncertainty. This approach is an example of the kind of "analytic deliberative process" that is widely acknowledged as the most effective way of dealing with complex environmental issues (Stern and Fineberg 1996; Dietz and Stern 1998; Dietz and Stern 2008).

This volume is one product of an ongoing analytic deliberative process that integrates these four components. We hope it can help readers think through the complex challenges of adapting to climate change, even as the processes that led to the volume have advanced the thinking of all of us involved.

Finally, we are pleased to report that the approach developed at the Symposium and described here is also the basis of the newly formed Great Lakes Integrated Sciences and Assessments Center (GLISA) (glisa.msu.edu; glisa.umich.edu). GLISA has been funded by the National Oceanographic and Atmospheric Administration to support climate adaptation in the region. Its goal is to provide decision makers in the region with better access to the best science, and to have the direction of scientific research informed by the needs of decision makers. In a very real sense, GLISA continues the work described in the chapter that follow.

REFERENCES

Allen Consulting Group. 2005. *Climate Change Risk and Vulnerability: Promoting an Efficient Adaptation Response in Australia.* Report to the Australian Greenhouse Office, Department of Environment and Heritage.

Arvai, J., G. Bridge, N. Dolsak et al. 2006. Adaptive management of the global climate problem: Bridging the gap between climate research and climate policy. *Climatic Change* 78(1):217–225.

Dessai, S., K. O'Brien, and M. Hulme. 2007. Editorial: On uncertainty and climate change. *Global Environmental Change* 17:1–3.

Dietz, T., and P.C. Stern. 1998. Science, values and biodiversity. *BioScience* 48:441–444.

———, eds. 2008. *Public Participation in Environmental Assessment and Decision Making.* Washington, DC: National Academy Press.

Fawcett, A.A., K.V. Calvin, F.C. de la Chesnaye, J.M. Reilly, and J.P. Weyant 2009. Overview of EMF 22 U.S. transition scenarios. *Energy Economics* 31:S198–S211.

Grothmann, T., and A. Patt. 2005. Adaptive capacity and human cognition: The process of individual adaptation to climate change. *Global Environmental Change* 15:199–213.

Janssen, M.A., and E. Ostrom. 2006. Editorial: Resilience, vulnerability, and adaptation: A cross-cutting theme of the International Human Dimensions Programme on Global Environmental Change. *Global Environmental Change* 16:237–239.

Karl, T.R., J.M. Melilo, and T.C. Peterson, eds. 2009. *Global Climate Change Impacts in the United States.* New York: Cambridge University Press.

Kasperson, J.X., R.E. Kasperson, and B.L. Turner II. 2010. Vulnerability of coupled human-ecological systems to global environmental change. In *Human Footprints on the Global Environment: Threats to Sustainability*, ed. by E. Rosa, A. Diekmann, T. Dietz, and C. Jaeger, 231–294. Cambridge, MA: The MIT Press.

Liu, J., T. Dietz, S.R. Carpenter et al. 2007a. Complexity of coupled human and natural systems. *Science* 317(5844):1513–1516.

———. 2007b. Coupled human and natural systems. *Ambio* 36(8):639–649.

Marx, S.M., E.U. Weber, B.S. Orlove et al. 2007. Communication and mental processes: Experiential and analytic processing of uncertain climate information. *Global Environmental Change* 17:47–58.

Natural Resources Canada. 2004. *Climate Change Impacts and Adaptation: A Canadian Perspective.* Ottawa: Climate Change Impacts and Adaptation Directorate, Natural Resources Canada.

Renn, O. 2008. *Risk Governance: Coping with Uncertainty in a Complex World.* London: Earthscan.

Rosa, E.A., A. Diekmann, T. Dietz, and C. Jaeger, eds. 2010. *Human Footprints on the Global Environment: Threats to Sustainability.* Cambridge MA: The MIT Press.

Smit, B., I. Burton, R. Klein, and J. Wandel. 2000. An anatomy of adaptation to climate change and variability. *Climatic Change* 45:223–251.

Smit, B., and J. Wandel. 2006. Adaptation, adaptive capacity and vulnerability. *Global Environmental Change* 16:282–292.

Solomon, S., G.-K. Plattner, R. Knutti, and P. Friedlingstein. 2009. Irreversible climate change due to carbon dioxide emissions. *Proceedings of the National Academy of Sciences* 106(6):1704–1709.

Stern, P.C., and H. Fineberg, eds. 1996. *Understanding Risk: Informing Decisions in a Democratic Society.* Washington, DC: National Academy Press.

U.S. National Research Council. 2010a. *Adapting to the Impacts of Climate Change.* Washington, DC: National Academies Press.

———. 2010b. *Advancing the Science of Climate Change.* Washington, DC: National Academies Press.

———. 2010c. *Climate Stabilization Targets: Emissions, Concentrations, and Impacts Over Decades to Millennia.* Washington, D.C.: National Academy Press.

———. 2010d. *Informing an Effective Response to Climate Change.* Washington, DC: National Academies Press.

———. 2010e. *Limiting the Magnitude of Climate Change.* Washington, DC: National Academies Press.

Vogel, C., S.C. Moser, R.E. Kasperson, and G.D. Dabelko. 2007. Linking vulnerability, adaptation, and resilience science to practice: Pathways, players, and partnerships. *Global Environmental Change* 17:349–364.

Willows, R., and R. Connell, eds. 2003. Climate adaptation: Risk, uncertainty, and decision-making. *United Kingdom Climate Impacts Programme.* Available at http://www.ukcip.org.uk/images/stories/Pub_pdfs/Risk.pdf.

PART ONE

Effects of Climate Change in the Great Lakes Region

Historical Climate Trends in Michigan and the Great Lakes Region

JEFFREY A. ANDRESEN

THE GREAT LAKES BASIN OF NORTH AMERICA CONTAINS THE LARG-est supply of freshwater in the world, with more than 20 percent of the global total (Quinn 1988) (figure 1). The region spans steep climate, geological, and vegetation gradients. Geological features transition from ancient crys-talline rocks of the continental craton overlain by glacial sediments in the

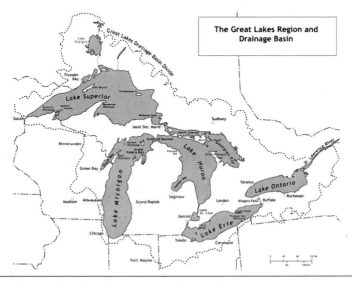

Figure 1. Geographical outline of the Great Lakes Basin of North America
Source: Courtesy of Schaetzl and Isard (2002).

17

north to a series of sedimentary rock strata covered by deep unconsolidated deposits in the south (Schaetzl and Isard 2002). Native vegetation also varies greatly, ranging from boreal forest in far northern sections to grassland along far southwestern fringes of the region.

THE CURRENT CLIMATE

The current climate of Michigan is chiefly governed by latitude, continental location, large-scale circulation patterns, and by the presence of the Great Lakes, which surround the state on three sides. The position of the polar jet stream generally controls day-to-day and week-to-week weather patterns in the winter and transition seasons, with somewhat less influence in the summer, when the state is also influenced by frequent incursions of warm, humid air masses of tropical origin (Andresen and Winkler 2009). In general, the presence of the Great Lakes tends to moderate air temperatures in Michigan relative to locations upstream of the prevailing westerly winds (e.g., Wisconsin, southwestern Ontario), largely through an increase in cloudiness. This moderation has important agricultural implications, especially in western areas of the state where hilly topography and the cold air drainage that results allow for an extensive tree fruit industry, including apples, cherries, and peaches.

The number and type of lake effect–related influences on weather and climate in Michigan and the Great Lakes region is large. Major associated climate impacts include seasonal increases or decreases in cloudiness, which in turn directly impacts insolation rates (the amount of sunlight reaching the ground) and air temperatures. In Michigan, weather is driven by the climatological source regions of relatively cold continental polar or arctic polar air masses in the interior sections of northern North America and the Arctic. This means a majority of lake-related cloudiness is associated with a northwesterly wind flow across the region during the fall and winter seasons. During the late spring and summer seasons, when lake water temperatures are relatively cooler than air and adjacent land surfaces, the impact on cloudiness is symmetrically opposite, as the cooler water leads to relatively greater atmospheric stability, general low-level atmospheric subsidence, and fewer clouds over and immediately downwind of the lakes. Air temperatures within Michigan are also directly affected by the moderating effects of air flow across the lakes, with a general reduction in temperatures in downwind areas during the spring and summer seasons, and an increase during the fall and winter seasons. Given enough atmospheric lift and moisture, clouds associated with wind flow over the lakes may also produce precipitation and altered precipitation patterns downwind of the lakes are among the most significant lake influences on regional climate. So-called lake effect snowfall greatly enhances the seasonal snowfall totals of areas generally within 100 kilometers of the downwind shores of the lakes.

Annual weather and climate cycles in Michigan and the Great Lakes region are linked directly to changes in the location of the polar jet stream, which in turn is linked to the annual cycle of solar insolation. Climatologically, the solar insolation cycle is closely related to the astronomical change of seasons, with some modifications associated with lake effect cloudiness. Solar insolation values tend to reach maximum levels in late June, with annual minimum values occurring from late December into early January. Mean temperatures in Michigan typically peak in late July or early August and reach minima during late January or early February. Coldest overall temperatures tend to be observed in interior areas away from the lakes. Total annual precipitation generally increases from minimum values just under 740 millimeters in the northeastern Lower Peninsula (climatologically the driest region in the USA east of the Mississippi River) to maximum levels of 950 millimeters or more in southwestern Lower Michigan. Total annual snowfall in the state is highly variable, but generally increases from minimum values of about 90 centimeters in the extreme southeastern Lower Peninsula to maximum values in the northwestern Lower and northern Upper Peninsulas, where annual mean totals in some locations exceed 550 centimeters. Seasonal snowfall totals and seasonal duration of snow cover there are climatologically among the greatest of any location in the USA east of the Rocky Mountains. In these major "snowbelt" areas (e.g., the spine of the Keweenaw Peninsula), a major portion of the snowfall is associated with lake effect processes (Changnon and Jones 1972), while over the remainder of the state, the majority of snowfall is due to larger, transient scale synoptic (large-scale) disturbances moving through the region.

PRE-INSTRUMENTAL TRENDS

Climate at a given location is defined as the "average meteorological conditions in a certain area over a certain period" (AMS 2000). Ideally, the search for climatological patterns and trends thus requires consistent, unbiased data from as many long-term sources as possible, as the magnitude of such trends may be far less than changes experienced on an annual, daily, or even hourly basis. In general, the amount and quality of data available for climatological analysis in the Great Lakes region decreases quickly with time into the past. Routine instrumental observations began in Michigan in the 1820s. However, because of the relatively low number and quality of those data, as well as differences in technology, historical data series reliable enough for use in climatological research typically do not begin until after 1880.

Fortunately, while not as detailed or precise as instrumental records, there are other, less direct methods of determining changes in climate. Based on fossil and other geological evidence, it is possible to demonstrate that climate in the Great Lakes region and many other regions of the world has varied markedly over geologic time scales, ranging from humid, tropical conditions during the Carboniferous and Devonian eras to frigid, glacial conditions as recently as 12,000 years ago, during the end of the Pleistocene era. These major shifts are thought to be the result of many factors, including tectonic drift of the continents, changes in the composition of the earth's atmosphere, periodic changes in the earth's tilt and orbit around the sun (Milankovitch cycles), and catastrophic singular events such as the impact of large meteorites and major volcanic eruptions.

More substantial paleoclimatological evidence of regional changes in climate is available since the end of the last major glacial epoch, or about 12,000 years before the present. Paleoclimatological data are typically gathered from sources such as pollen and other limnological indicator counts as well as element isotope concentrations. During early portions of the Holocene era approximately 10,000 years before the present (YBP), climate in the region warmed rapidly following the end of the last major glacial epoch, resulting in a relatively mild and dry climate (versus current and recent past conditions) that lasted until about 5,000 YBP. During this period, the levels of the Great Lakes fell until the lakes became terminal or confined about 7,900 YBP (Croley and Lewis 2006), and vegetation in the region gradually transitioned from a dominance of boreal (northern cold weather) to xeric (dry habitat) species (Webb et al. 1993). Beginning about 5,000 YBP, climate cooled and precipitation totals increased, possibly associated with a change in jet-stream patterns across North America from mostly west to east, or zonal flow to more north–south or meridional flow (Wright 1992). The cooler, wetter climate favored the establishment of more mesic (moderately moist habitat) vegetation, which is among the primary vegetation types today. Given a more meridional jet-stream flow (and an increase in frequency of polar and arctic-origin air masses into the region), there is also evidence to suggest that the frequency and amount of lake effect precipitation increased relative to previous periods at about 3,000 YBP (Delcourt et al. 2002). Finally, during the late Holocene, the Great Lakes region experienced a period of relatively mild temperatures from approximately 800 C.E. to 1300 C.E. (sometimes referred to as the "Medieval Warm Period"), followed by a period of relatively cool temperatures from about 1400 C.E. until the late nineteenth century (the "Little Ice Age"). Thermometer-derived observations of temperature were taken in the region during the very end of the latter period, and many extreme minimum-temperature records were set that still stand today.

CLIMATE TRENDS SINCE THE LATE EIGHTEENTH CENTURY

Even with instrumental observation records of climate, the search for trends is complicated, as significant changes in climatological behavior can be introduced artificially by changes in station location, station environment, observation time, observer, and type of instrument (e.g., Karl and Williams 1987; Andresen and Dale 1986). In this paper, we consider data taken from the NOAA Climatological Cooperative Network (NOAA/NCDC 1895–2006), which were quality controlled and archived by the Michigan state climatologist in East Lansing.

Temperature Trends

For obtaining general temperature trends for the state, monthly temperature data averaged across all ten climatological divisions in Michigan from 1895–2008 are presented in figure 2. In this figure and many subsequent figures, a nine-year moving average was calculated from the yearly data and plotted on the graph (drawn as the thicker line) to illustrate longer-term, roughly decadal patterns. For all moving average values, the number plotted represents the mean of a nine-year period centered on the year (e.g., a value plotted for 1905 represents an average over the period 1901–1909). From the figure, there are some discernible temperature patterns, including a period of steady temperatures from about 1900–1930, a slow cooling trend of about 0.5°C from 1930–1980, and a more rapid warming trend from 1980 (about 1.3°C). Across the entire 112-year record, the overall change is about 0.6°C. The trends and temporal patterns are somewhat similar to overall global trends, which include an increase in mean temperature of about 0.8°C since 1850 (IPCC 2007).

The increases in temperature during the past 25 years were found not

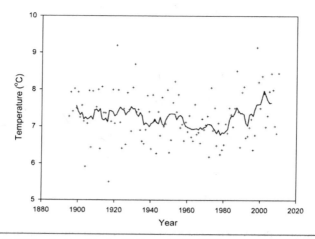

Figure 2. Mean average temperature for Michigan, 1895–2008
Note: Individual years are plotted as individual tick marks and the 9-year moving average is plotted as a solid line.

to be consistent across season or time of day. A relatively greater proportion of the regional warming appears to be associated with warmer nighttime temperatures (i.e., minimum temperatures), which is in agreement with the earlier findings of Lorenz et al. (2009a) and Easterling et al. (1997). Just as importantly, much of the warming occurred during the winter and spring seasons, with relatively little change in temperature found during the summer and fall seasons. The latter results are consistent with the results of Zhang et al. (2001), who found that the largest increases in temperature across southern Canada between 1900 and 1998 had occurred in winter and early spring. To illustrate this trend, mean winter minimum temperatures are plotted versus time in figure 3 for Ironwood, Michigan, a representative site in the western Upper Peninsula. The increase in winter temperatures beginning about 1980 coincides almost exactly with the increase in statewide mean temperatures, and the magnitude of warming of winter minimum temperatures during the last 25 years of record is about 2.0°C. Among the impacts resulting from the recent warmer winter temperatures is a reduction in the amount and duration of ice cover on the Great Lakes. This is well illustrated in figures 4 and 5, which depict time trends of the frequency of ice cover on the western arm of Grand Traverse Bay since 1851, and the number of days each winter that more than 20 percent of the Great Lakes were covered in ice.

The data in figure 4 are unique due to the record length and completeness. They clearly suggest that ice formation on this water body has been

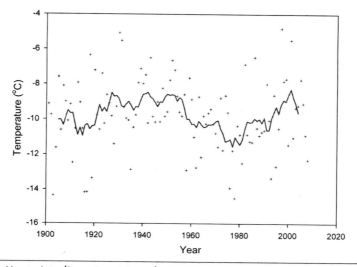

Figure 3. Mean winter (December–February) temperatures at Ironwood, MI, 1961–2009
Note: Individual years are plotted as individual tick marks and the 9-year moving average is plotted as a solid line.

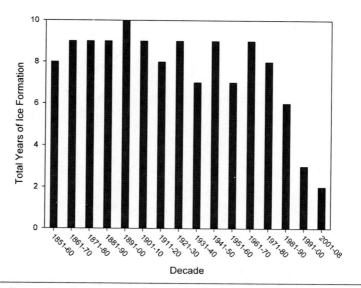

Figure 4. The number of years per decade that ice has covered the western arm of Grand Traverse Bay in northwestern Lower Michigan, 1851–2008
Source: Data courtesy of Mr. James Nugent, Suttons Bay, MI, and the Traverse City Chamber of Commerce, Traverse City, MI.

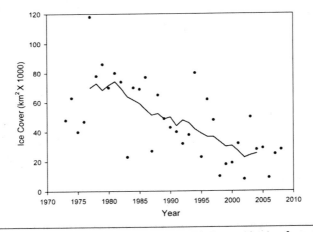

Figure 5. The number of days each winter season in which the combined surfaces of the Great Lakes were more than 20 percent ice covered, 1973–2008
Source: Data courtesy of the NOAA Great Lakes Environmental Research Laboratory, Ann Arbor, MI.

more common in the past, with an average of 7 to 9 years per decade from the beginning of the record in 1851 through about 1980. Since then, the number has fallen off dramatically, with ice formation in only 2 of the past 8 years. While available for a much shorter period of record, satellite imagery provides a more comprehensive estimate of ice cover on the lakes.

In figure 5, the number of days appears to peak during the late 1970s, falling to an average of less than 20 days per season during the last several years of record. These numbers are in good agreement with the results of Duguay et al. (2006), who documented similar decreases in ice-cover duration as well as trends towards earlier lake ice breakup in the spring season during the period 1951–2000 in nearby areas in Canada.

Precipitation Trends

Historical precipitation trends in the region are illustrated in figure 6, which depicts state-level annual precipitation totals versus time averaged across all ten of Michigan's climatological divisions. Precipitation generally decreased during the first 35 years of the series (1895–1930), followed by a general increasing trend beginning during the late 1930s that has persisted to the present. During the increase of the past several decades, there was a notable period of relatively dry conditions during the late 1950s and early 1960s. There is

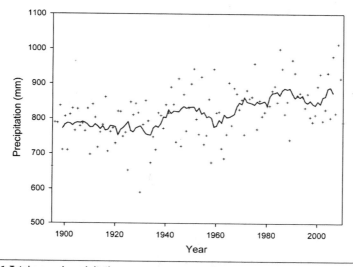

Figure 6. Total annual precipitation versus year averaged across Michigan, 1895–2008
Note: Individual years are plotted as individual tick marks and the 9-year moving average is plotted as a solid line.

also evidence of at least a temporary leveling-off in the upward precipitation trend, with drier conditions observed during the last five to six years. In general, annual precipitation has varied by decade, with totals ranging from approximately 650 to 950 millimeters. Overall, the 1930s can be seen as the driest decade on record, while the 1980s were arguably the wettest, which is consistent with records across the region (Lorenz at al. 2009b).

From the 1930s through the present, annual precipitation increased by approximately 1.5 millimeters. It is interesting to note that this increase has not occurred as the result of an increase in the number of heavy precipitation events, as illustrated by the frequency of 25-millimeter precipitation events over time at East Lansing, Michigan (figure 7). While somewhat variable over time, these heavy events were found to generally remain within a range of one to eight events per year. Statewide increases in precipitation over time were instead found to be associated with significant overall increases in the number of wet days, and the number of wet days following wet days. A representative time series of single and two-day consecutive wet-day frequencies from Caro, Michigan, is given in figure 8. The upward trends in both single and two-day consecutive wet-day frequencies are impressive, with increases of more than 30 percent between the 1930s and the present.

Given more annual precipitation and days with precipitation over time,

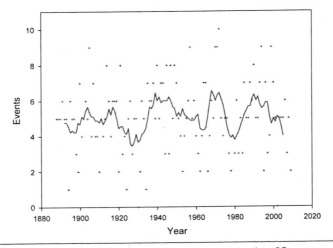

Figure 7. Total annual number of daily precipitation events greater than 25mm versus year for East Lansing, MI, 1895–2008
Note: Individual years are plotted as individual tick marks and the 9-year moving average is plotted as a solid line.

it is also logical to assume that cloudiness in the region has increased as well. Unfortunately, solar-radiation observational records in the state and region are scarce. Cloudiness data obtained from hourly aviation observations at Lansing, MI, are plotted versus time in figure 9. With the exception of a period during the late 1980s, they suggest that cloudiness increased over time from the 1960s through the 1990s, which is consistent with the increases in precipitation observed during the same time frame.

Snowfall

As noted earlier, snow in the Great Lakes region is generally associated either with large, synoptic-scale weather disturbances or with the lake effect phenomenon, which may lead to highly varying snowfall totals over only short distances. Seasonal snowfall totals at two representative sites in Michigan illustrate two contrasting trends (figures 10 and 11). In the first, at Bay City, Michigan, a site generally outside of the lake effect snowbelts, seasonal snowfall from 1920–2008 is seen to increase slightly over the entire record period, but generally decrease from the late 1950s through 2005. At Chatham, Michigan, which is within an area of the central Upper Peninsula frequented by lake effect snowfall, the trend is very different. Seasonal snowfall decreased slightly during the 1920s and 1930s, but increased significantly

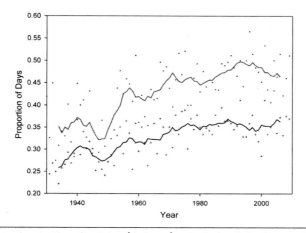

Figure 8. Annual proportion of wet days (dark gray) and wet days following wet days (light gray) at Caro, Michigan, 1930–2008

Note: Nine-year moving averages are plotted as solid lines.

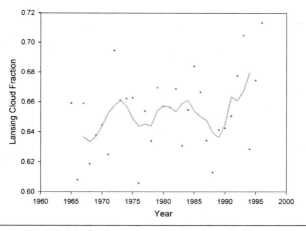

Figure 9. Average annual cloud fraction versus year at Lansing, MI, 1960–1996 based on hourly aviation observations

Note: Individual years are plotted as individual tick marks and the 9-year moving average is plotted as a solid line.

during the 1940s into the 1990s. There is evidence of a leveling-off, or even a decreasing trend during the last decade of the series.

While these two series from sites only a few hundred kilometers apart appear contradictory, they represent the two basic trend patterns found across the state and region, and the differences are likely associated with some of the trends described earlier. The milder wintertime temperatures of

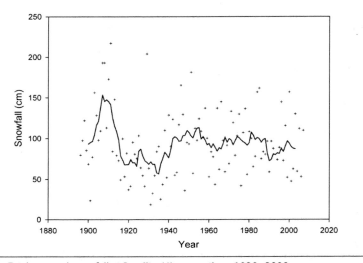

Figure 10. Total seasonal snowfall at Bay City, MI, versus time, 1920–2008
Note: Individual years are plotted as individual tick marks and the 9-year moving average is plotted as a solid line.

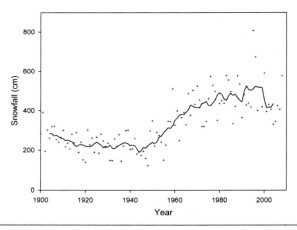

Figure 11. Total seasonal snowfall at Chatham, MI, versus time, 1920–2008
Note: Individual years are plotted as individual tick marks and the 9-year moving average is plotted as a solid line.

the past few decades have resulted in a reduction of ice cover on the Great Lakes, which in turn has led to more open water and milder lake temperatures, both of which are positively associated with lake effect precipitation. Lake effect snowfall is also strongly dependent on the passage of cold polar or arctic air masses into the region from the high latitudes of the Northern

Hemisphere. Whether or not the recent winter warming has had an impact on this aspect of lake effect precipitation (note the snowfall decreases at Chatham since the 1990s) is not clear. This pattern has also been documented by Burnett et al. (2003) across the Lower Great Lakes as well.

Finally, it is also interesting to consider the amount and duration of snow cover in a given season. In general, data series from most climatological observation sites in the state suggest that the amount of snow cover has decreased with time during the past few decades (data not shown). In figure 12, the number of days each winter season with 2.5 centimeters or more snow cover is plotted versus year for the Chatham, Michigan, site. Even at a site where seasonal snowfall has almost doubled in the last half century, the number of days with snow cover decreased by almost 30 percent beginning around 1980. While the number has rebounded somewhat in recent years, the data suggest that the milder winter temperatures are melting snow more quickly than in past decades, even though more snow is falling.

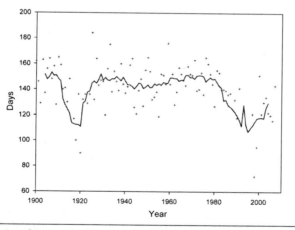

Figure 12. Number of days each winter season with 2.5 cm of snow cover or greater, Chatham, MI, 1920–2008

Note: Individual years are plotted as individual tick marks and the 9-year moving average is plotted as a solid line.

IMPACT OF CLIMATIC TRENDS ON AGRICULTURE:
AN ILLUSTRATION WITH CORN PRODUCTION

As an example of the potential impacts of a changing climate on agriculture, a process-based crop simulation model, CERES-Maize (Jones and Kiniry 1986), was utilized to simulate corn production at Coldwater, Michigan, under historical conditions from 1900–2006. All agronomic and soil inputs in the simulations were held constant to isolate the impacts of weather and climate. Agronomic inputs included crop growth-and-yield-potential coefficients of a typical current corn hybrid grown in the Great Lakes region, which were taken from the results of Andresen et al. (2001). Soil profiles used in the simulation were Marlette loam (fine-loamy, mixed mesic Haplic Glossudalf), which is representative of a fine-textured Michigan agricultural soil. Planting dates were set at May 1st (calendar day 121) for all simulations.

   Time series of simulated yield and total growing-season precipitation versus year from these simulations are plotted in figure 13. In the figure, the correlation between the simulated yields and seasonal precipitation is clearly evident. Both variables decrease from relatively high values during the first decades of the twentieth century to a low during the 1930s, followed by an increase again through the end of the record period. From the same simulations, yields from the mid-1930s through 2008 were found to increase at an average rate of 8.1 kilograms per hectare each year. Similarly, increases in soil moisture available to the plant at mid-season and the amount of plant available water in the soil profile across the entire

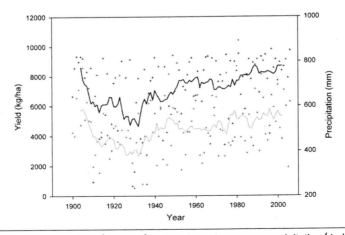

Figure 13. Simulated corn yields (light gray) and total growing-season precipitation (dark gray) versus time at Coldwater, MI, 1900–2008

season, both key variables in determining ultimate yield potential, were also found to increase at rates of 0.15 millimeters and 0.19 millimeters per year, respectively. It is thus likely that some of the regional gains in agronomic productivity during the past half century have been due to wetter, less stressful growing-season conditions.

Finally, it is interesting to note that recent climate projections suggest a general continuation of trends toward a warmer and wetter regional climate in the future (Christensen et al. 2007). Potential positive impacts under this scenario include continued increases in agronomic productivity due to improved climatic conditions as well as increased CO_2 levels (Andresen et al. 2000). Negative impacts include potential increases in insect, weed, and disease pressure (Easterling et al. 2007).

If the magnitude of regional climate changes in the future reaches levels suggested by many climate-model projections (see the Mackey and Winkler et al. chapters in this volume for more details), farmers will be forced to adapt to the changes or become uncompetitive and unprofitable (e.g., see Bradshaw et al. this volume). Future adaptations could include longer season or different crop varieties, double cropping systems, use of irrigation, and other unforeseen technological improvements. Due to the rates of climate change, it is possible that growers may have to rely on much shorter time horizons for long-term planning than were utilized in past years (e.g., 5 to 10 years versus 30 or 40 years). There is evidence based on past history that agriculture could at least partially adapt to a changing climate, and that the costs of such adaptations would be small compared to costs associated with an expansion of, or changes to, major production areas (Doering et al. 1997). Ultimately, the ability to adapt will depend upon the nature of the climatic change, as increases in variability could make future adaptations difficult (Easterling et al. 2007).

SUMMARY

Climate across the Great Lakes region has varied markedly during the past, ranging from tropical to glacial and including almost everything in between. Instrumental measurement records of the most recent century suggest some general temporal trends across the region. Mean temperatures warmed from approximately 1900 to 1940, followed by a cooling trend from the early 1940s to the late 1970s, followed by a second warming trend that began around 1980 through the present. Much of the warming trend during the past two or three decades has been associated with warmer minimum temperatures during the winter and spring seasons. Following an abnormally dry decade during the 1930s, the region has also progressively become wetter. Seasonal snowfall has trended upwards in areas of the region frequented by lake effect snowfall, while totals have remained steady or decreased slightly in non–lake effect areas.

Collectively, the warmer and wetter climate of the past few decades has likely resulted in increases of some regional crop yields and is consistent with projections of future climate in the region.

REFERENCES

American Meteorological Society (AMS). 2000. *Glossary of Meteorology*. 2nd ed. Boston: American Meteorological Society.

Andresen, J.A., G. Alagarswamy, J.T. Ritchie, C.A. Rotz, and A.W. LeBaron. 2001. Assessment of the impact of weather on maize, soybean, and alfalfa production in the Upper Great Lakes Region of the United States, 1895–1996. *Agronomy Journal* 93:1059–1070.

Andresen, J.A., G. Alagarswamy, D. Stead, W.B. Sea, and H.H. Cheng. 2000. Agriculture. In *Preparing for a Changing Climate: The Potential Consequences of Climate Variability and Change*, ed. by P.J. Sousounis and J.M. Bisanz. Ann Arbor: University of Michigan, Atmospheric, Oceanic, and Space Sciences Department.

Andresen, J.A., and R.F. Dale. 1986. Climatic data bases. *Proceedings of the Indiana Academy of Science* 95:503–512.

Andresen, J.A. and J.A. Winkler. 2009. Weather and climate. In *Michigan Geography and Geology*, ed. by R.J. Schaetzl, D. Brandt, and J.T. Darden. Boston: Pearson Custom Publishing.

Burnett A.W., M.E. Kirby, H.T. Mullins, and W.P. Patterson. 2003. Increasing Great Lake–effect snowfall during the twentieth century: A regional response to global warming? *Journal of Climate* 16:3535–3542.

Changnon, S.A., and D.M.A. Jones. 1972. Review of the influences of the Great Lakes on weather. *Water Resources Research* 8:360–371.

Christensen, J.H., et al. 2007. Regional climate projections. In *Climate Change 2007: The Physical Science Basis*, ed. by S. Solomon et al. Cambridge: Cambridge University Press.

Croley, T.E., and C.F.M. Lewis. 2006. Warmer and drier climates that make terminal great lakes. *Journal of Great Lakes Research* 32:852–869.

Delcourt, P.A., P.L. Nester, H.R. Delcourt, C.I. Mora, and K.H. Orvis. 2002.

Holocene lake-effect precipitation in Northern Michigan. *Quarternary Research* 57:225–233.

Doering, O., et al. 1997. Mitigation strategies and unforeseen consequences: A systematic assessment of adaptation of Upper Midwest agriculture to future climate change. *World Resource Review* 9:447–459.

Duguay, C.R., T.D. Prowse, B.R. Bonsal, R.D. Brown, M.P. Lacroix, and P. Menard. 2006. Recent trends in Canadian lake ice cover. *Hydrological Processes* 20:781–801.

Easterling, W.E., et al. 2007. Food, fibre and forest products. In *Climate Change 2007: Impacts, Adaptation and Vulnerability*, ed. by M.L. Parry et al. Cambridge: Cambridge University Press.

Easterling, D.R., B. Horton, P.D. Jones et al. 1997: Maximum and minimum temperature trends for the globe. *Science* 277:364–367.

Intergovernmental Panel on Climate Change (IPCC). 2007. *Summary for Policymakers from Climate Change 2007: The Physical Science Basis.* Geneva: IPCC Secretariat, c/o World Meteorological Organization. Available at http://www .ipcc.ch/SPM2feb07.pdf.

Jones, C.A., and J.R. Kiniry. 1986. *CERES-Maize: A Simulation Model of Maize Growth and Development.* College Station: Texas A&M University Press.

Karl, T.R., and C.N. Williams. 1987. An approach to adjusting climatological time series for discontinuous inhomogeneities. *Journal of Climate and Applied Meteorology* 26:1744–1763.

Lorenz, D.J., S.J. Vavrus, D.J. Vimont et al. 2009a. Wisconsin's changing climate: Temperature. Chapter 7 in *Regional Climate Variability and Change in the Midwest*, ed. by S.C. Pryor. Bloomington, IN: Midwest Assessment Group for Investigations of Climate, Indiana University Press.

Lorenz, D.J., S.J. Vavrus, D.J. Vimont et al. 2009b. Wisconsin's changing climate: Hydrologic cycle. In *Regional Climate Variability and Change in the Midwest*, ed. by S.C. Pryor, 135–144. Bloomington: Indiana University Press.

NOAA/NCDC. 1895–2006. Climatological Data. National Oceanic and Atmospheric Administration, National Climatic Data Center, Asheville, NC.

Quinn, F.H. 1988. Great Lakes water levels, past, present, and future. In "The Great Lakes: Living with North America's Inland Waters." *Proceedings of the American Water Resources Association Annual Meeting, Milwaukee, WI*, 83–92.

Schaetzl, R.J., and S.A. Isard. 2002. The Great Lakes Basin. Chapter 16 in *The Physical Geography of North America*, ed. by A.R. Orme. New York: Oxford University Press.

Webb, T., P.J. Bartlein, S.P. Harrison, and K.H. Anderson. 1993. Vegetation, lake levels, and climate in eastern North America for the past 18,000 years. In *Global*

Climates Since the Last Glacial Maximum, ed. by H.E. Wright, J.E. Kutzbach, T. Webb et al., 415–467. Minneapolis: University of Minnesota Press.

Wright, H.E. 1992. Patterns of Holocene climatic change in the Midwestern United States. *Quarternary Research* 38:129–134.

Zhang, X., W.D. Hogg, and E. Mekis. 2001. Spatial and temporal characteristics of heavy precipitation events over Canada. *Journal of Climate* 14:1923–1936.

Climate Change Impacts and Adaptation Strategies for Great Lakes Nearshore and Coastal Systems

SCUDDER D. MACKEY

WITHIN THE GREAT LAKES BASIN, INDIVIDUAL SPECIES, BIOLOGICAL communities, and the ecosystem as a whole have adapted to a natural range of physical and environmental conditions that are controlled by the interaction of master variables—climate, geology, and hydrology. Climate change has the potential to significantly alter the physical integrity of the Great Lakes by altering the natural processes and pathways that convey energy, water, and materials through the basin with impacts to sustainable water resources, habitat, biodiversity, and ecological function. Ecological responses will be driven primarily by changes to physical characteristics of the environment.

MASTER VARIABLES

Master variables are fundamental characteristics that structure, organize, and define a system, influence the distribution and abundance of energy and materials, and regulate processes that have a profound effect on physical, chemical, and biological integrity and ecosystem function. When altered or changed, the effects of these master variables cascade through physical, chemical, and biological systems altering processes and ecosystem function.

There are six master variables, and each of those master variables are linked to specific system integrity components.

The first three are natural variables that structure, organize, and define the fundamental physical and energy characteristics of the landscape and the processes that act on that landscape. The second three are anthropogenic variables that impact the structure and organization of the landscape and the processes that act on that landscape—but are directly linked to anthropogenic activities from within, or outside, the Great Lakes Basin.

It is important to recognize that there are attributes of these master variables that cannot be manipulated and are therefore not actionable. Examples would include climate (temperature, precipitation, solar flux); geology (bedrock and surficial materials); or regional basin geomorphology. However, other attributes are actionable and can be altered to obtain a desired result.

Examples of actionable attributes would include hydrology (flow regime, flow paths and hydraulic connectivity, diversions, breaching of watersheds); chemical pollution (pollutant and nutrient loadings); biological pollution (introduction, dispersion, and establishment of invasive species); or resource utilization and extraction (changes in land cover, water diversions, consumptive uses). By focusing on these master variables and working to

Natural Variables	Integrity Components
Climate (energy)	**Physical**
Geology (materials, soils, geomorphology, bathymetry)	**Physical, Chemical**
Hydrology (water quantity, quality, surface and groundwater flow, hydrography)	**Physical, Chemical**
Anthropogenic Variables	
Chemical Pollution (what enters the system)	**Chemical**
Biological Pollution (what enters the system)	**Biological**
Resource Utilization (what is anthropogenically removed, consumed, or altered within the system)	**Physical, Chemical, Biological**

Figure 1. Master variables for nearshore and coastal systems. Master variables linked to fundamental integrity components. Natural variables define the fundamental physical and energy characteristics of the system. Anthropogenic variables affect the structure and organization of the landscape and processes that act on the landscape.
Source: Mackey (2008).

restore them to a more natural condition, we allow *natural system processes* to maintain and restore essential ecosystem functions over the long term with minimal anthropogenic management (or interference). This approach to restoration is both economically and ecologically efficient (Mackey 2008).

PHYSICAL INTEGRITY

Great Lakes habitats are inextricably linked to physical integrity. Habitat is the critical component that links biological communities and ecosystems to natural processes, pathways, and the landscape. Physical integrity is achieved when the physical components of a system, and the natural processes and pathways that structure, organize, define, and regulate them, correspond to undisturbed natural conditions and are mutually supportive and sustainable (Mackey 2008). Sustainable natural processes are created when master variables interact to convey energy, water, and materials through a system in ways that correspond to undisturbed natural conditions, maintain system integrity, and promote system resiliency and regeneration—irrespective of natural and anthropogenic perturbations. Maintaining physical integrity implies that master variables—the fundamental factors that structure, organize, and define a system; influence the distribution and abundance of energy and materials; and regulate processes—are functioning in a naturalized state (Mackey 2008, 11): "Physical integrity is achieved when the physical components of a system and the natural processes and pathways that structure, organize, define, and regulate them correspond to undisturbed natural conditions and are mutually supportive and sustainable." This does not imply that undisturbed conditions (those that existed before human settlement) are a prerequisite to achieve physical integrity. Physical integrity can be achieved by duplicating natural conditions and functions in ways that maintain ecosystem health and promote ecosystem resiliency and regeneration. This approach allows the ecosystem to adjust and adapt irrespective of natural and anthropogenic perturbations, including the potential long-term effects of climate change.

The pattern and distribution of habitats are controlled, in part, by the underlying physical characteristics of the basin and interactions between energy, water, and the landscape (e.g., Sly and Busch 1992; Higgins et al. 1998; Mackey and Goforth 2005; Mackey 2008). Moreover, the physical characteristics and energy conditions that define habitats are created by the interaction of master variables—climate (energy), geology (geomorphology and substrate), and hydrology (water mass characteristics and flow)—the same variables and processes that maintain physical integrity. Habitats are created when there is an intersection of a range of physical, chemical, and biological characteristics that meet the life-stage requirements of an organism.

From the perspective of physical integrity, *physical habitat* is defined by a range of physical characteristics and energy conditions that can be delineated geographically and that meet the needs of a specific species, biological community, or ecological function for a given life stage (Mackey 2008). To be utilized as habitat, these physical characteristics and energy conditions must exhibit an organizational pattern, persist, and be "repeatable"—elements that are essential to maintain a sustainable and renewable resource (Peters and Cross 1992). The repeatable nature of habitat implies that the natural processes that create physical habitat must also be repeatable and persist over a range of spatial and temporal scales.

For example, seasonal changes in flow, thermal structure, and water-mass characteristics create repeatable patterns and connections within tributaries and lakes within the basin. Spatially these patterns occur within the same general locations year after year. Moreover, the pattern of movement of water, energy, and materials through the system (which depends on connectivity) also exhibits an organizational pattern, persists, and is repeatable. These patterns and connections, in part, control the seasonal usage of Great Lakes fish spawning and nursery habitats (Chubb and Liston 1985).

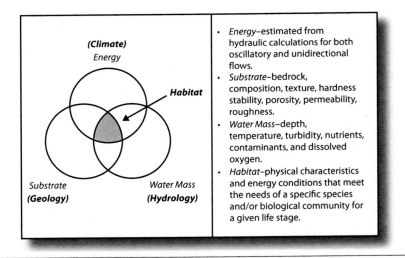

Figure 2. Fundamental components of physical habitat in aquatic systems
Source: Mackey (2008).

HABITAT INTEGRITY

It therefore follows that the quality and integrity of Great Lakes habitats are maintained by sustainable natural processes, pathways, and landscapes that persist and are repeatable over a range of spatial and temporal scales. *Habitat integrity* is created by protecting and restoring sustainable natural processes, pathways, and landscapes that maintain a range of physical, chemical, and biological characteristics and energy conditions that can be delineated geographically and that meet the needs of a specific species, biological community, or ecological function for a given life stage (Mackey 2008). The importance of sustainable natural processes and pathways to ecological function cannot be overemphasized.

A convenient and useful way to view the Great Lakes is by defining environmental zones based on dominant processes and/or hydrogeomorphic characteristics (Mackey et al. 2006). Environmental zones represent a combination of abiotic conditions that are associated with dominant processes and/or hydrogeomorphic characteristics, but are *not* linked to any particular species, biological community, or ecological function. The rationale for using this approach is based on the need to identify critical characteristics and processes that control the distribution, pattern, and function of aquatic habitats irrespective of ecological function. Moreover, environmental zone-specific indicator suites can be developed and tailored to monitor and assess critical parameters within each environmental zone.

For example, high-quality fish and aquatic habitats are created by a unique set of environmental conditions and processes that together meet the life-stage requirements of a species, biological community, or ecological function (Mackey 2008). These processes play a significant role, ultimately determining the distribution and utilization of essential fisheries habitats within the Great Lakes system. The environmental zone concept provides a logical way to partition the Great Lakes system based on dominant natural processes and zone-specific hydrogeomorphic characteristics.

To summarize, the distribution, pattern, and function of aquatic habitats within the Great Lakes are controlled by underlying hydrogeomorphic characteristics and the dominant physical processes that form them. Moreover, the processes acting within environmental zones are distinct and yield a unique response from associated biological communities. Climate-induced changes to natural processes, pathways, and landscapes will significantly affect Great Lakes coastal-margin and nearshore habitats.

CLIMATE CHANGE AND ECOLOGICAL INTEGRITY

Recent research and modeling results suggest that anticipated long-term changes in climate have the potential to significantly alter the physical integrity of the Great Lakes Basin (e.g., Kling et al. 2003; Solomon et al. 2007). Changes in climate may be gradual and will be affected by interactions between natural long-term climatic cycles and potential long-term impacts due to anthropogenic changes to the earth's atmosphere. Because climate and hydrology are master variables, these changes are likely to have a significant impact not only on physical integrity, but the chemical, biological, and ecological integrity of the Great Lakes as well.

Details of climate-change predictions resulting from climate-modeling efforts are described elsewhere and are beyond the scope of this work (e.g., Mortsch and Quinn 1996; Lee et al. 1996; Magnuson et al. 1997; Mortsch 1998; Casselman 2002; Lofgren et al. 2002; Brandt et al. 2002; Wuebbles and Hayhoe 2004; Kling et al. 2003; Mortsch et al. 2006; Solomon et al. 2007). However, climate-change-induced alterations to weather, i.e., precipitation, evapotranspiration, and storm frequency, severity, and patterns, will likely alter several aspects of the physical and habitat integrity of the Great Lakes Basin. For example:

- *Tributary and groundwater flows, base flows:* seasonal alterations of flow regime; spatial and temporal shifts in seasonal timing
- *Great Lakes water levels:* a general lowering of water levels; loss of connectivity; altered coastal-margin and nearshore habitat structure
- *Thermal effects:* thermal stratification; altered open-lake and nearshore surface water temperatures, circulation patterns, and processes; reduced ice cover; spatial and temporal shifts in seasonal timing
- *Latitudinal shifts in ecoregions:* regional changes in land and vegetative cover and associated terrestrial and aquatic communities and habitats

Specifically, climate change may directly affect the processes and pathways that control the volume, timing, and movement of water through riverine, coastal-margin, and nearshore systems, including (1) spatial and temporal changes in flow and water-level regime (water depth, velocity, flow patterns); (2) altered connections and pathways within the land-water interface; and (3) thermal effects (temperature). The impact of these changes will be first observed in coastal-margin areas where changes in flow and lake-level regimes will change shoreline locations and directly impact connectivity and alter hydroperiods in shallow-water and coastal-margin environments.

Initially these effects will be localized, but will become cumulative through time. Altered weather patterns and changing water levels will modify the energy distribution and nearshore coastal processes that shape and maintain Great Lakes shorelines and nearshore habitats.

The effects of climate change in coastal areas will be exacerbated by anthropogenic activities, especially in areas where submerged lands may be exposed and pressure to develop coastal areas is high. Filling of shoreline areas for development, increased shoreline armoring (shore protection), and dredging and removal of submergent and emergent aquatic vegetation to promote water access will have a detrimental effect on Great Lakes nearshore and coastal-margin environments. Moreover, as average temperatures increase and water availability becomes more critical, coastal areas of the Great Lakes will be subject to increasing development pressure. Unfortunately, additional work is needed to assess how the combined effects of climate change and development pressure will alter long-term stressor gradients in Great Lakes coastal-margin and nearshore areas.

GREAT LAKES WATER LEVELS

Within Great Lakes coastal-margin and open-water systems, the equivalent of natural flow regime is the natural water-level regime. Great Lakes water-level regimes are controlled primarily by the interaction of two master variables, climate and hydrology. Water levels represent the integrated sum of water inputs and losses from the system—typically expressed by a hydrologic water-balance equation—that are driven by climate (long-term and seasonal weather patterns), hydrology, and flow regime (surface water, groundwater, and connecting channel flows), and utilization of water resources within the basin (water withdrawals, diversions, and connecting channel flows) (Quinn 2002). Climatic controls, including precipitation, evapotranspiration, and the frequency, duration, and pattern of major storm events are typically driven by seasonal and longer-term climatic cycles (Quinn 2002; Baedke and Thompson 2000). Long-term and seasonal changes in precipitation and evaporation result in the interannual and seasonal variability of water levels and associated connecting channel flows within, and between, all of the Great Lakes (Derecki 1985; Lenters 2001; Quinn 2002).

Change in lake water levels can be characterized in ways similar to flow regimes, where the fundamental characteristics of flow—magnitude, frequency, timing, duration, and rate of change—can also be applied to Great Lakes water levels and connecting channel flows (Poff et al. 1997). Also influencing the water-level regimes are short-term fluctuations in water levels that are caused, in part, by local wind or storm events that perturb the water surface that may not necessarily reflect a change in the overall water balance of the lake or basin under consideration (seiche is an oscillatory change in the water level surface due to wind or storm events). These short-term fluctuations in water level may also have important structuring effects on coastal-margin and open-lake ecosystems.

LAKE ST. CLAIR—A SENTINEL BASIN

For the purpose of examining potential climate change impacts, Lake St. Clair is a shallow connecting lake fed by the mouth of the St. Clair River that drains Lake Huron, which in turn flows into the Detroit River, which empties into Lake Erie (figure 3). Because it is so shallow and is responsive due to a very short residence time of water in the lake, Lake St. Clair serves as an ideal sentinel basin for long-term climate change effects that may occur in the Great Lakes.

Regional climate change models (Canadian Centre for Climate Modeling

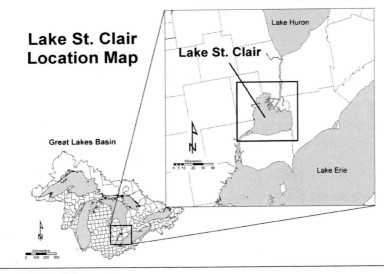

Figure 3. Lake St. Clair

CCGM1 and UKMO/Hadley Centre HADCM2) project more than a 1 to 1.5 meter decline in long-term annual water levels over the next 70 years for the Great Lakes (e.g. Lee et al. 1996; Mortsch and Quinn 1996; Lofgren et al. 2002; Sousounis and Grover 2002). More recent work by Wuebbles and Hayhoe (2004) using the HADCM3 model projects higher temperature changes for the Midwestern United States than those predicted by the CCGM1 and HADCM2 models. Lee et al. (1996) predicted a reduction in long-term annual water levels in Lake Erie and Lake St. Clair by more than a meter.

Recent work by David Fay and Yin Fan (Environment Canada) is summarized in figure 4 for Lake St. Clair (modified from Mortsch et al. 2006). This table provides a range of predicted water levels for four different climate change scenarios on Lake St. Clair. The base case represents current climatic and water-level conditions. Even though there is a degree of uncertainty in these analyses, in all cases these models predict a general decline in Lake St. Clair annual water levels, ranging from 0.2 meters to as much as 1 meter within the next 50 years. Similar declines are predicted for Lakes Michigan-Huron and Lake Erie. However, more recent downscaled models suggest water levels near the long-term mean and/or present-day water levels. Note that these models are predicting mean annual water levels, and that the range of variability about these means is expected to increase in three of the four cases listed in figure 4.

However, these water-level scenarios do not take into account long-term cyclicity in Great Lakes water levels (Rob Nairn, W.F. Baird & Associates, Oakville, ON, pers. communication). Extrapolating prehistoric Great Lakes water-level changes using cores and radiometric dating, long-term water-level cycles with periods of approximately 160 years, 60 years, 33 years, and 4 to 8 years have been identified over the past 4,700 years for the Great Lakes (e.g., Baedke and Thompson 2000; Baedke et al. 2004) (figure 5). These cycles generally correspond to changes in historic water levels even though there is considerable variability. However, the extrapolated long-term cyclic changes in water level *should not be used to predict absolute changes in water level,* but are used to illustrate general long-term cyclic trends that must be considered when evaluating potential impacts of climate change.

Superimposed on these daily, seasonal, and longer-term climatic cycles is the potential for long-term climate change. Long-term projections for Lake St. Clair suggest that water levels will remain near current levels for the next 15 to 20 years and then, under this scenario, may decline precipitously in response to climate change impacts superimposed on cyclic long-term lake levels (figure 6). Note that the anticipated long-term water levels are

	Base Case	Warm & Dry (CGCN2 A21)	Not as Warm & Dry (CGCM2 B23)	Warm & Wet (HadCM3 A1F1)	Not as Warm & Dry (HadCM3 B22)
LakeStatistics					
Mean	175.38	174.40	174.75	174.57	175.18
Maximum	176.11	175.12	175.43	175.36	175.95
Minimum	174.38	173.37	173.72	173.46	174.05
Annual Range	1.73	1.75	1.71	1.90	1.90
Change from Base Case					
Annual		-0.98	-0.63	-0.81	-0.20
Winter		-0.95	-0.62	-0.81	-0.21
Spring		-0.98	-0.61	-0.77	-0.16
Summer		-1.01	-0.64	-0.80	-0.20
Fall		-1.01	-0.65	-0.87	-0.26
Growing Season		-1.01	-0.63	-0.78	-0.18

Figure 4. Predicted Lake St. Clair water levels, in meters (IGLD 1985 datum). Predicted changes in Lake St. Clair water levels for four climate change scenarios within the next 50 years. Even though most models indicate somewhat lower future water levels, there are some model scenarios where water levels remain near present-day water levels.
Source: David Fay and Yin Fan, Environment Canada—modified from Mortsch et al. (2006).

presented as a range of values that correspond to values for change from base case in figure 4. The range of anticipated water levels reflects the predictive uncertainty associated with statistically generated long-term cyclic changes in water levels, and the even greater uncertainty associated with long-term climate change predictions. This is highlighted by the fact that there are some model scenarios where water levels are expected to remain near the long-term mean and/or present-day water levels.

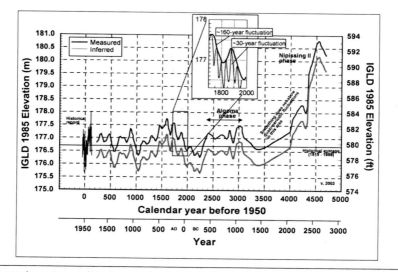

Figure 5. Long-term cyclicity in water levels in Lake Michigan-Huron
Source: Courtesy of Dr. Todd Thompson, Indiana Geological Survey.

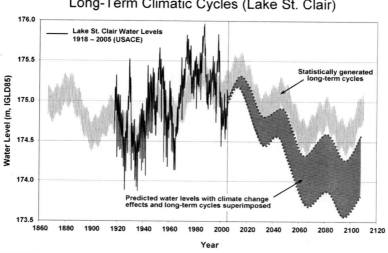

Figure 6. Long-term climatic cycles in Lake St. Clair. Future Lake St. Clair water levels may be influenced by potential climate change effects superimposed on long-term climatic cycles. Light gray band represents superimposed statistically generated long-term 160-year, 33-year, and 4- to 8-year long-term cycles for Lake St. Clair (as described by Baedke and Thompson 2000). Given modeling and statistical uncertainties, it is also plausible that water levels may remain near the long-term mean and/or present-day water levels.
Source: Modified from workshop presentation—Rob Nairn, W.F. Baird & Associates (2005).

WATER-LEVEL REGIME IMPACTS

Even though the predicted range of future water levels falls generally within historical limits, the lows recorded in the mid 1920s and 1930s lasted for relatively short periods of time and are considered to be extreme events. Under this scenario, model predictions suggest that mean water levels will be substantially lower than current levels for extended periods of time, perhaps decades. Coastal-margin and nearshore habitats are created by the interaction of water-level regimes, natural coastal processes, open-lake circulation processes and patterns, and the connections and pathways along which these processes act. Climate-induced reductions in water levels will "reset" how biological communities utilize energy, materials, and water as it is conveyed through coastal-margin and open-lake systems. Moreover, rapid phenological changes related to changes in water-level regimes may occur as many aquatic species have adapted to seasonal fluctuations in temperature and water level (Root and Hall, this volume).

Recent work by Burkett et al. (2005) suggests that when ecological threshold values are exceeded, ecosystem responses may be rapid and nonlinear. In other words, relatively minor physical changes may result in significant ecological effects. Moreover, these effects have the potential to cascade through the system, disrupting established ecological processes and functions. These changes will result in irreversible long-term changes in the distribution of coastal-margin and nearshore aquatic habitats and ecological function.

For example, anticipated reductions in Lake St. Clair water levels by up to a meter over the next 50 years will significantly alter the location of the Lake St. Clair shoreline, particularly in shallow-water areas with gentle slopes (e.g., Edsal and Cleland 1989; Lee et al. 1996; Kling et al. 2003; Mackey et al. 2006). Based on updated bathymetry (figure 7), up to 20 percent of the present surface area of Lake St. Clair may eventually be exposed and converted into land with a 1 meter drop in lake level (figure 8). In some areas, the Lake St. Clair shoreline may be extended lakeward by as much as 6 kilometers (e.g., Lee et al. 1996; Mackey et al. 2006). As a result, the location of nearshore littoral and sublittoral habitats will change substantially, affecting benthic and fish community distribution throughout the entire system (littoral means shallow-water areas affected by waves). Somewhat less dramatic changes would occur along other Great Lakes shorelines in response to climate-induced lowering of water levels as well.

Of significance is the fact that the broad, shallow platform created by the St. Clair Delta will be exposed, reducing the currently available area

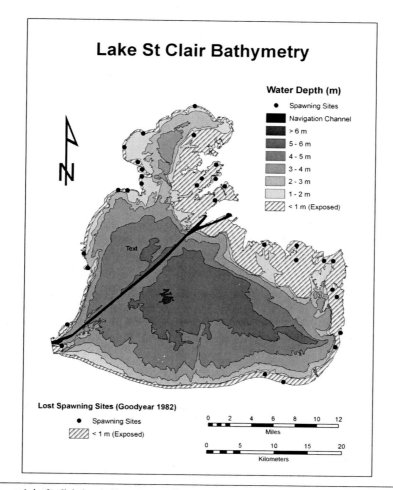

Figure 7. Lake St. Clair 1 m bathymetry and fish-spawning locations reported by Goodyear et al. (1982). A 1 m drop in lake level may expose up to 22,000 ha (54,000 ac or 85 sq mi) of the lakebed by 2050. Critical fish spawning and nursery habitats will shift lakeward as water levels decline. Polygons were derived from bathymetry provided by the National Geophysical Data Center (1998). Water depths are referenced to Low Water Datum (174.4 m).
Source: GLFC Lake Erie GIS—University of Michigan Institute for Fisheries Research (Mackey et al. 2006).

of shallow-water habitat (less than 1 meter water depth) by more than 50 percent (figure 8). The surface area of Lake St. Clair that is currently 1 to 2 meters deep comprises only 9,231 hectares. Even if this area was completely suitable as "newly available" shallow-water habitat, it would constitute only 40 percent of the preexisting area, which would represent a major loss of

	Area (ha)	*Cumulative Area (ha)*	*Percent (%)*
0 to 1	22,690	22,690	20.2
1 to 2	9,231	31,922	8.2
2 to 3	11,480	43,401	10.2
3 to 4	20,067	63,468	17.8
4 to 5	28,587	92,055	25.4
5 to 6	19,004	111,059	16.9
> 6	206	111,265	0.2
Navigation Channel	1,161	112,426	1.0
Total Area		112,426	100.0

Figure 8. Lake St. Clair hypsography, based on a water surface level of 174.4 meters (Low Water Datum). More than 50 percent of current shallow-water habitats will be lost if Lake St. Clair water levels drop by 1 meter, excluding additional reductions that may result due to loss of wetland connectivity.
Source: Mackey et al. (2006).

shallow-water habitats available for colonization. Moreover, nearshore slopes adjacent to the new shoreline will be steeper, allowing more wave energy to impinge directly on the shoreline. Fine-grained sediments will be eroded and transported offshore, with remaining coarse-grained sediments redistributed into narrow beaches, bars, barriers, and spits. Dewatering and compaction of sediments may also cause gradual subsidence of the delta platform.

Climate-induced reductions in water levels will also hydrologically isolate many high-quality wetland and estuarine areas (e.g., Mortsch 1998; Wilcox et al. 2002; Wilcox 2004; Mortsch et al. 2006). Moreover, many of these wetland and estuarine areas provide critical spawning, nursery, and forage habitat for Great Lakes fish communities (Goodyear et al. 1982). Over 40 percent of Lake Erie's fish species are classified as wetland-dependent or facultative wetland-dependent. Altered flow paths and shifts in seasonal timing will affect the reproductive behavior of the fish that have adapted to the natural water level regime and flows (e.g., Edsall and Cleland 1989; Casselman 2002; Kling et al. 2003). In summary, repositioning of the shoreline will significantly alter environmental conditions and the distribution of coastal-margin and nearshore habitats within the nearshore and coastal-margin areas of Lake St. Clair.

Within Lake St. Clair, there are 43 sites used by 33 fish species identified in Goodyear et al. (1982). Of the 43 sites, 28 are located in shallow water

(water depths less than 1 meter) and would likely be dewatered and exposed as Lake St. Clair water levels decline (figure 7). Figure 9 lists the total number of sites per species and includes a general description of preferred spawning habitat and temperature for individual species. The combined effects of habitat loss, loss of connectivity, and altered temperature regimes may adversely affect species that spawn or rely on nursery habitats in coastal wetland or shallow-water environments.

THERMAL IMPACTS

The effects of lower water levels will not be limited to habitat, but may fundamentally affect seasonal timing and connectivity, food web dynamics, and the distribution, structure, composition, and abundance of fish communities in Lake Erie, Lake St. Clair, and associated connecting channels (Casselman 2002; Kling et al. 2003; Root and Hall this volume). For example, a decrease in total water volume could lead to greater spatial overlap between predators and prey and could elevate predator-prey interactions thereby possibly increasing predation and/or feeding rates for some species at certain life stages (GLFC 2005).

A warmer climate and lower water levels (e.g., Sousounis and Grover 2002) could also affect the thermal structure of the Great Lakes, causing changes in both lake chemistry and lake ecology. Climatic effects, such as lower precipitation and higher evaporation rates, could result in lower water levels, higher surface-water temperatures, a deeper and stronger thermocline, and in Lake Erie, a reduced water volume in the hypolimnion (more dense and cold waters below the thermocline) resulting in more frequent episodes of central basin anoxia, i.e., the Lake Erie "Dead Zone." Reduced water volumes in the hypolimnion, combined with selective feeding and altered nutrient cycling by invasive zebra/quagga mussels (Dreissenid spp.), may increase decomposition rates and thereby rapidly deplete dissolved oxygen in the hypolimnion (Lam et al. 1987, 2002; Charlton and Milne 2004). Warmer water temperatures, a deeper thermocline, and an expanded "dead zone" will cause shifts in the distribution of both cold- and warm-water fishes in Lake Erie and its tributaries.

Warmer water temperatures may yield an increase in potential secondary production rates and nutrient recycling, coupled with an increase in the prey demand required to meet those higher rates (Shuter and Meisner 1992; Shuter and Ing 1997; Brandt et al. 2002). Moreover, the rate

Species	Sites	Environment	Temperature	Country
Smallmouth bass	24	Intermediate	warm	US, Canada
Muskellunge	17	Intermediate	warm	US, Canada
Yellow perch	15	Intermediate	cool	US, Canada
Northern pike	14	Intermediate	cool	US, Canada
Largemouth bass	13	Coastal wetland	warm	US, Canada
Bluegill	11	Coastal wetland	warm	US, Canada
Carp	7	Intermediate	warm	US, Canada
Channel catfish	7	Intermediate	warm	US, Canada
Emerald shiner	7	Intermediate	cool	US, Canada
Rock bass	7	Coastal wetland	cool	US, Canada
Walleye	7	Intermediate	cool	US, Canada
Lake sturgeon	6	Open water	cool/cold	US, Canada
Pumpkinseed	6	Coastal wetland	warm	US, Canada
Spottail shiner	6	Intermediate	cold/cool	US, Canada
Common shiner	5	Coastal wetland	cool	US, Canada
Spotfin shiner	5	Coastal wetland	warm	US, Canada
Black crappie	4	Intermediate	cool	US, Canada
Golden shiner	4	Coastal wetland	cool	US, Canada
Longnose gar	4	Coastal wetland	warm	US, Canada
Alewife	3	Intermediate	cold	US, Canada
Goldfish	3	Coastal wetland	warm	US, Canada
White crappie	3	Coastal wetland	cool	US, Canada
Brown bullhead	2	Coastal wetland	warm	US, Canada
Lake whitefish	2	Open water	cold	US, Canada
Rainbow smelt	2	Intermediate	cold	US, Canada
Sea lamprey	2	Open water	cold	US, Canada
White bass	2	Coastal wetland	warm	US, Canada
Bowfin	1	Coastal wetland	warm	Canada
Gizzard shad	1	Coastal wetland	cool	Canada
Logperch	1	Coastal wetland	cool/warm	Canada
Trout-perch	1	Intermediate	cold	Canada
White sucker	1	Intermediate	cool	Canada

Figure 9. Reported Lake St. Clair spawning locations (Goodyear et al. 1982). Summary list of species that use spawning and/or nursery sites potentially affected by long-term reductions in Lake St. Clair water levels (Goodyear et al. 1982). Individual sites may be used by multiple species during different times of the year. Species highlighted in gray may be more severely affected by changes in water-level regime and temperature than other species.
Source: GLFC Lake Erie GIS, University of Michigan Institute for Fisheries Research (Mackey et al. 2006).

of contaminant (mercury) accumulation in the food chain may increase due to increases in temperature and the relationship between temperature and mercury contamination in fish (Bodaly et al. 1993; Yediler and Jacobs 1995). Increases in water temperature are positively correlated with mercury methylation rates and increase the availability of methyl mercury for incorporation into fish tissue.

The abundance of important recreational and commercial fish species (lake trout, walleye, northern pike, and lake whitefish) varies with the amount of thermally suitable habitat (Christie and Regier 1988; Lester et al. 2004). A warm thermal structure may cause a northward shift of 19 boundaries for both warm- and cold-water fishes, affecting abundance, distribution, and sensitivity to exploitation (Minns and Moore 1992; McCormick and Fahnenstiel 1999; Casselman 2002; Kling et al. 2003). Warming could also remove thermal constraints that have protected the Great Lakes in the past, and increase the potential number of organisms that can successfully invade the lake (Mandrak 1989). Moreover, in response to shifted thermal boundaries, zebra/quagga mussels, round gobies, and other aquatic nuisance species could expand their existing ranges northward into the upper Great Lakes as well (GLFC 2005).

DISCUSSION

Global climate circulation models have been used to predict changes in temperature, weather, precipitation, storm severity and frequency, and indirectly Great Lakes water levels. These predictions have a high degree of uncertainty and represent a range of possible futures or scenarios. For all of these scenarios, the physical integrity of the Great Lakes will be modified or altered in response to changing climate conditions. Thus, ecological responses to climate change will be driven primarily by changes in physical integrity, and these responses may be nonlinear, especially if boundaries and thresholds are exceeded (Burkett et al. 2005). Moreover, these responses may cascade through the entire system making long-term prediction of long-term ecological impacts difficult.

However, within the Great Lakes Basin an important, but often overlooked, fundamental fact is that native fish and aquatic communities have co-evolved and adapted to the physical characteristics of the system—including the distribution, pattern, and function of aquatic habitats and the timing and seasonality of the dynamic processes that create and maintain those habitats (e.g., Busch and Lary 1996; Jones et al. 1996). Actions taken to restore natural processes, pathways, and landscapes will result in a positive response by the ecosystem, and over time, will yield long-term benefits including sustainable water resources and improvements in habitat, biodiversity, and ecological function. More importantly, by restoring natural processes, pathways, and landscapes, inherent natural structuring processes will increase ecosystem resiliency and allow the ecosystem to more readily adapt to changing environmental conditions. It is only by maintaining sustainable processes that we will be able to maintain a sustainable and resilient ecosystem (Mackey 2008).

Conservation and natural-resource management agencies have long recognized the potential consequences of altered thermal and water-level regimes due to climate change, but have not sufficiently incorporated the effects of climate change into long-term conservation or management plans (e.g., The Nature Conservancy 2000; Rodriguez and Reid 2001). As a result, these plans do not provide for the future conservation of coastal and submerged nearshore areas where new wetlands, coastal embayments, and high-quality fish habitats will form (e.g., Saxon 2003). Nor do current planning efforts incorporate the potential effects of altered climate, flow, and thermal regimes on watersheds, tributaries, nearshore and coastal-margin areas, or the Great Lakes themselves. The reason for this may be due, in part, to inherent uncertainties associated with climate change predictions and natural ecological complexity. Moreover, the ecological tools and models necessary to generate these predictions are only now being developed and are in their infancy.

Given that climate change will first and foremost affect physical integrity, and that changes to physical integrity will drive ecological change, it would seem logical to better understand biophysical linkages between organisms, communities, and ecological functions within a physical habitat context; and then apply physical and hydrodynamic models to predict climate-induced change in coastal-margin and nearshore habitats. Moreover, our ability to monitor and model physical components of the system is much more robust than our ability to monitor and model biological or ecological components of the system.

For example, to guide future management decisions, it is essential to understand how lower long-term water levels will change basin and connecting-channel configurations and alter the distribution of fish and aquatic habitats within the Great Lakes system. This is particularly important for highly productive shallow areas of the basin, especially embayments and nearshore areas. In response to changes in water level, coastal-margin and nearshore aquatic communities will respond and adapt to physical changes in the environment. However, the mechanisms of habitat change are poorly understood. Traditional views of aquatic habitat succession are typically two-dimensional. However, aquatic habitats are three-dimensional, and habitat succession will also necessarily be three-dimensional in response to declining water levels (figure 10).

In response to water-level change, these "step-stone" or transitional habitats may shift sequentially (contiguous succession) or shift spontaneously (noncontiguous succession) in response to suitable climate-influenced abiotic factors (figure 10) (Saxon 2003). Transitional habitats represent refugia, or areas of relative habitat stability that can be used by existing biological communities during periods of climate-driven change. Transitional habitats shift through time and space in response to changing abiotic conditions. Integrated physical models can be used to predict the timing, location, and distribution of these transitional habitats, and may be used to infer the extent of dominant climate-driven change processes—displacement and/or fragmentation—and the critical climate-related thresholds likely to cause them.

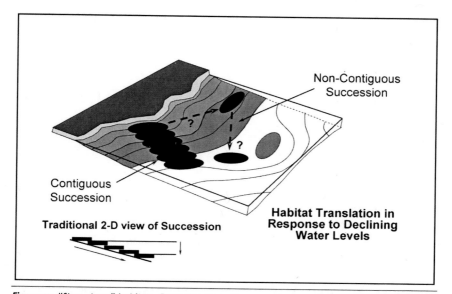

Figure 10. "Step-stone" habitats. Examples of contiguous and noncontiguous succession of aquatic habitat in response to declining water levels. "Step-stone" or transitional habitats (Saxon 2003) represent refugia or areas of relative habitat stability during periods of climate-driven change and are necessarily three-dimensional.

Within coastal-margin and nearshore zones, natural coastal processes include oscillatory and unidirectional flows generated by waves and currents, with the resulting conveyance of material and energy along the shore and between the coastal margin and the open lake. These processes control the distribution of materials and substrates in the coastal-margin (water depths 0 to 3 meters) and nearshore zones (water depths 3 meters to 15 meters).

Figure 11 is an example of how heuristic physical models could be used to predict the hydrogeomorphic characteristics of a portion of the Lake St. Clair shoreline in response to a 1 meter drop in Lake St. Clair water levels. Within the St. Clair Delta and other Great Lakes tributaries, a lowering of water levels will also cause increased erosion and channel deepening as tributary channels equilibrate to a lower base level. The main channels will incise into underlying sediments, and eroded materials will be deposited in deeper water areas located offshore from tributary mouths. However, coarse-grained sediments may be transported into the coastal-margin zone by littoral currents and form protective beaches and barriers along an actively eroding low-relief shoreline. This scenario is based, in part, on an understanding of how natural coastal processes may alter the shoreline to create new protected embayments and shallow-water habitats. An understanding of transitional or "step-stone" habitats and biophysical linkages between physical habitat and associated biological communities is also necessary to generate an appropriate ecological response to this scenario.

Extensive coastal engineering works and loss of littoral sand from adjacent coastal-margin and nearshore areas have created habitats that are now much more coarse-grained and heterogeneous than would have naturally been present along many Great Lakes coastlines. It is anticipated that as Great Lakes water levels decline, littoral sand deposits will become stranded at higher shoreline elevations and lost to the active littoral system (M. Chrzastowski, Illinois DNR, pers. communication, 2006). The loss of these sand resources may be significant, especially along sand-poor Great Lakes cohesive shorelines.

One of the consequences of these substrate changes is the rapid colonization and spread of aquatic invasive species (such as Dreissenid spp.) that have adversely impacted food web dynamics and the Great Lakes ecosystem. It is only now recognized that many of the physical changes that have occurred in the nearshore zones of the Great Lakes have provided the opportunity for massive expansion of these invasive species, along with significant associated ecological impacts (e.g., Janssen et al. 2004, Meadows et al. 2005).

Even though there are significant uncertainties associated with the

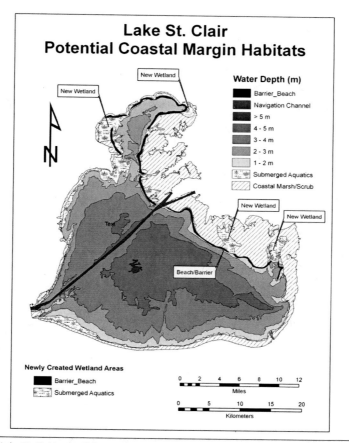

Figure 11. Lake St. Clair potential coastal-margin habitats. Example of how the Lake St. Clair coastline may change in response to a 1 meter drop in water levels. Once water levels stabilize, beaches will form and the movement of sand will create protective barriers and spits at embayment mouths. The inflows from the eastern distributary channels will be diminished and locally sourced. Shallow-water areas and protected embayments will have both submergent and emergent aquatic vegetation.
Source: Mackey et al. (2006).

predictions illustrated in figure 11, there is a reasonable probability that the area identified will develop ecologically important attributes that are worthy of protection. Even though our current political system requires a high degree of certainty before taking action, it may be acceptable to take action in the face of uncertainty as long as the actions taken have a reasonable probability of success. Moreover, there is nothing to lose by building and validating these heuristic models and tools, because if the "best case" climate-change scenario turns out to be reality, the models can be used to more effectively protect and manage the resource. If the "worst-case" climate-change scenario becomes reality, then the models can be used to provide guidance as to what policies need to be developed and where they should be implemented to minimize anthropogenic interference as the ecosystem adjusts and adapts to a new set of environmental conditions. By focusing on the physical and chemical processes that create and sustain critical aquatic and fish habitats, it is possible to directly quantify the degree of habitat improvement without having to measure (or model) the more complex biological components or functions of the ecosystem.

RECOMMENDATIONS

Additional work is needed to more fully understand the biophysical linkages between physical habitats, associated biological communities, and the natural processes and pathways that connect them. Future changes to the ecosystem may yield changes that have not yet been observed and for which data do not exist. It is only through an understanding of biophysical processes that we may be able to predict the ecological responses of the Great Lakes ecosystem due to changes in water-level regime. Moreover, additional tools/models need to be developed that integrate physical and ecological processes to simulate potential changes in environmental conditions and associated aquatic habitats resulting from long-term changes in water-level regime. Using these models, it will be possible to heuristically identify potential long-term management, protection, and restoration opportunities based, in part, on an understanding of biophysical processes.

The resulting management, conservation, and protection strategies must be designed to protect potential refugia, transitional, and newly created coastal-margin and nearshore habitat areas from anthropogenic modification and/or degradation. As water levels recede, there will be increasing societal pressure to develop and modify newly exposed areas of the shoreline. Critical reaches of the Great Lakes shoreline (as identified by the long-term models) must be protected and preserved to ensure that essential ecological functions are maintained during periods of transition.

It will also be necessary to establish a long-term, aquatic-habitat research and monitoring effort within the Great Lakes to track changes and continually update and refine the heuristic models. An important consideration will be to identify the appropriate variables to be monitored, and to establish thresholds or triggers that tell us *when* to modify resource management and protection policies. Scheraga (this volume) highlights the critical need for a robust decision-support system that uses these data to provide guidance at multiple scales across the Great Lakes basin. This approach will provide the knowledge and science-based tools to build the capacity of key agencies, organizations, and institutions to identify and implement sustainable protection, restoration, and enhancement opportunities.

This discussion highlights the need to incorporate management and research strategies designed to address uncertainty and respond to potential long-term stressors, such as climate change, water diversions, and continued growth and development that have the potential to impair the physical integrity of the Great Lakes. Moser (this volume) describes how scientific uncertainty can be incorporated into decision-making processes, and recommends several strategies that could be used to assess and manage climate-change impacts. Given the uncertainties associated with climate change, it is necessary to implement a proactive anticipatory management approach (commonly referred to as adaptive management strategies) that identifies long-term planning, protection, and restoration needs in response to climate-change induced stressors and impairments within the Great Lakes Basin. Application of adaptive management strategies will help to ensure the physical and ecological integrity of the Great Lakes in the face of major environmental change.

SUMMARY

Protection and restoration of natural processes, pathways, and landscapes will improve the resiliency and regenerative capacity of physical and biological systems to resist potential long-term natural and anthropogenic stressors including the effects of climate change. Maintaining a sustainable ecosystem requires a "balanced" approach to ecosystem protection and restoration—an approach that includes consideration of sustainable natural processes, pathways, and landscapes as part of a comprehensive protection and adaptation strategy for the Great Lakes in the face of ongoing climate change.

NOTE

The author would like to thank Dr. Jan Ciborowski (University of Windsor), Dr. John Gannon (IJC), Phil Ryan (OMNR-Retired), Dr. Susan Doka (DFO), Dr. Rob Nairn (W.F. Baird & Associates), Michelle DePhilip (TNC), and members of the Lake Erie Habitat Task Group (GLFC) for their contributions and insightful discussion on physical integrity, coastal-margin and nearshore aquatic habitats, lake levels and flow regimes, and climate-change issues in the Great Lakes. Funding for portions of this work came from the International Joint Commission Windsor Regional Office and the US Fish and Wildlife Service Great Lakes Fisheries Restoration Act, Agreement #30181–4-J259.

REFERENCES

Baedke, S.J., and T.A. Thompson. 2000. A 4,700-year record of lake level and isostasy for Lake Michigan. *Journal of Great Lakes Research* 26(4):416–426.

Baedke, S.J., T.A. Thompson, J.W. Johnston, and D.A. Wilcox. 2004. Reconstructing paleo lake levels from relict shorelines along the upper Great Lakes. *Aquatic Ecosystem Health and Management* 7(4):1–15.

Bodaly, R.A., J.W.M. Rudd, R.J.P. Fudge, and C.A. Kelly. 1993. Mercury concentrations in fish related to size of remote Canadian shield lakes. *Canadian Journal of Fisheries and Aquatic Sciences* 50:980–987.

Brandt, S., D. Mason, M. McCormick, B. Lofgren, and T. Hunter. 2002. Climate change: Implications for fish growth performance in the Great Lakes. *American Fisheries Society Symposium* 32:61–76.

Burkett, V.R., D.A. Wilcox, R. Stottlemyer et al. 2005. Nonlinear dynamics in ecosystem response to climatic change: Case studies and policy implications. *Ecological Complexity* 2(4):357–394.

Busch, W.D.N., S.J. Lary. 1996. Assessment of habitat impairments impacting the aquatic resources of Lake Ontario. *Canadian Journal of Fisheries and Aquatic Sciences* 53(suppl. 1):113–120.

Casselman, J.M. 2002. Effects of temperatures, global extremes, and climate change on year-class production of warmwater, coolwater and coldwater fishes in the Great Lakes basin. *American Fisheries Society Symposium* 32:39–60.

Charlton, M.N., and J.E. Milne. 2004. *Review of Thirty Years of Change in Lake Erie Water Quality*. NWRI Contribution No. 04-167, Burlington, ON.

Christie, G.C., and H.A. Regier. 1988. Measures of optimal thermal habitat and their relationship to yields for four commercial fish species. *Canadian Journal of Fisheries and Aquatic Sciences* 45:301–314.

Chubb, S., and C.R. Liston. 1985. Relationships of water level fluctuations and fish. In *Coastal Wetlands*, ed. by H.H. Prince and F.M. D'Itri, 121–140. Chelsea, MI: Lewis Publishers, Inc.

Derecki, J.A. 1985. Effect of channel changes in the St. Clair River during the present century. *Journal of Great Lakes Research* 11(3):201–207.

Edsall, T., and J. Cleland. 1989. *Effects of Altered Water Levels and Flows on Fish in the Great Lakes Connecting Channels*. IJC Water Levels Reference Study, Task Group 2, Task 202–4 Report. USFWS, National Fisheries Research Center, Ann Arbor, MI.

Goodyear, C.D., T.A. Edsall, D.M. Ormsby-Dempsey, G.D. Moss, and P.E. Polanski. 1982. *Atlas of Spawning and Nursery Areas of Great Lakes Fishes*. USFWS, Report FWS/OBS-82/52, vols. 1–14, Washington, DC.

Great Lakes Fisheries Commission (GLFC). 2005. *Lake Erie Environmental Objectives*. Report of the Environmental Objectives Subcommittee of the Lake Erie Committee, Great Lakes Fisheries Commission, July 2005.

Higgins, J., M. Lammert, M. Bryer, M. DePhilip, and D. Grossman. 1998. Freshwater conservation in the Great Lakes basin: Development and application of an aquatic community classification framework. Chicago: The Nature Conservancy, Great Lakes Program.

Janssen, J., M.B. Berg, and S.J. Lozano. 2004. Submerged terra incognita: Lake Michigan's abundant but unknown rocky zones. In *The State of Lake Michigan: Ecology, Health, and Management*, ed. by T. Edsall and M. Munawar. Stuttgart, Germany: Ecovision World Monograph Series, Aquatic Ecosystem Health and Management Society.

Jones, M.L., R.G. Randall, D. Hayes et al. 1996. Assessing the ecological effects of habitat change: Moving beyond productive capacity. *Canadian Journal of Fisheries and Aquatic Sciences* 53(suppl. 1):446–457.

Kling, G.W., K. Hayhoe, L.B. Johnson et al. 2003. *Confronting Climate Change in the Great Lakes Region: Impacts on our Communities and Ecosystems*. Cambridge, MA: Union of Concerned Scientists; Washington, DC: Ecological Society of America.

Lam, D.C.L., W.M. Schertzer, and A.S. Fraser. 1987. A post-audit analysis of the NWRI nine-box water quality model for Lake Erie. *Journal of Great Lakes Research* 13:782–800.

Lam, D.C.L., W.M. Schertzer, and R.C. McCrimmon. 2002. Modelling changes in phosphorus and dissolved oxygen pre- and post-zebra mussel arrival in Lake Erie. NWRI Contribution No. 02–198, Environment Canada, Burlington, ON.

Lee, D.H., R. Moulton, and D.A. Hibner. 1996. Climate change impacts on Western Lake Erie, Detroit River, and Lake St. Clair water levels: Great Lakes–St. Lawrence Basin Project, Environment Canada and NOAA, GLERL Contribution #985.

Lenters, J.D. 2001. Long-term trends in the seasonal cycle of Great Lakes water levels. *Journal of Great Lakes Research* 27(3):342–353.

Lester, N.P., A.J. Dextrase, R.S. Kushneriuk, M.R. Rawson, and P.A. Ryan. 2004. Light and temperature: key factors affecting walleye abundance and production. *Transactions of the American Fisheries Society* 133:588–605.

Lofgren, B.M., F.H. Quinn, A.H. Clites, R.A. Assel, A.J. Eberhardt, and C.L. Luukkonen. 2002. Evaluation of potential impacts on Great Lakes Water Resources based on climate scenarios of two GCM's. *Journal of Great Lakes Research* 28(4):537–554.

Mackey, S.D. 2008. *Physical Integrity of the Great Lakes: Opportunities for Ecosystem Restoration.* Report to the Great Lakes Water Quality Board, International Joint Commission, Windsor, ON.

Mackey, S.D., and R.R. Goforth. 2005. Great Lakes nearshore habitat science. In "Great Lakes nearshore and coastal habitats," ed. by S.D. Mackey and R.R. Goforth. Special Issue, *Journal of Great Lakes Research* 31 (suppl. 1):1–5.

Mackey, S.D., J.M. Reutter, J.J.H. Ciborowski, R.C. Haas, M.N. Charlton, and R.J. Kreis. 2006. *Huron-Erie Corridor System Habitat Assessment—Changing Water Levels and Effects of Global Climate Change.* Project Completion Report, USFWS Restoration Act Sponsored Research, Agreement #30181–4-J259.

Magnuson, J.J., K.E. Webster, R.A. Assel et al. 1997. Potential effects of climate change on aquatic systems: Laurentian Great Lakes and Precambrian Shield Region. *Hydrological Processes* 11(6).

Mandrak, N.E. 1989. Potential invasion of the Great Lakes by fish species associated with climate warming. *Journal of Great Lakes Research* 15:306–316.

McCormick, M.J., and G.L. Fahnenstiel. 1999. Recent climatic trends in nearshore water temperatures in the St. Lawrence Great Lakes. *Limnology and Oceanography* 44:530–540.

Meadows, G.A., Mackey, S.D., Goforth, R.R. et al. 2005. Cumulative impacts of nearshore engineering. In "Great Lakes nearshore and coastal habitats," ed. by S.D. Mackey and R.R. Goforth. Special Issue, *Journal of Great Lakes Research* 31(suppl. 1):90–112.

Minns, C.K., and J.E. Moore. 1992. Predicting the impact of climate change on the spatial pattern of freshwater fish yield capability in eastern Canadian lakes. *Climatic Change* 22:327–346.

Mortsch, L.D. 1998. Assessing the impact of climate change on the Great Lakes shoreline wetlands. *Climate Change* 40:391–416.

Mortsch, L.D., E. Snell, and J. Ingram. 2006. Climate variability and changes within the context of the Great Lakes basin. Chapter 2 in *Great Lakes Coastal Wetland Communities: Vulnerability to Climate Change and Response to Adaptation Strategies*, ed. by L. Mortsch, J. Ingram, A. Hebb, and S. Doka, 9–19. Toronto: Environment Canada and the Department of Fisheries and Oceans.

Mortsch, L.D., and F.H. Quinn. 1996. Climate change scenarios for Great Lakes Basin ecosystem studies. *Limnology and Oceanography* 41:903–911.

National Geophysical Data Center. 1998. Bathymetry of Lake Erie and Lake Saint Clair. In *World Data Center for Marine Geology and Geophysics Report #MGG-13*, ed. by L.A. Taylor, P. Vincent, and J.S. Warren. Boulder, CO: National Geophysical Data Center.

Nature Conservancy, The. 2000. A conservation blueprint for the Great Lakes. Chicago: The Nature Conservancy, Great Lakes Program.

Poff, N.L., J.D. Allan, M.B. Bain et al. 1997. The natural flow regime: A paradigm for river conservation and restoration. *BioScience* 47:769–784.

Peters, D.S., and F.A. Cross. 1992. What is coastal fish habitat? In *Stemming the Tide of Coastal Fish Habitat Loss*, ed. by Richard H. Stroud, 17–22. Proceedings of a Symposium on Conservation of Coastal Fish Habitat, Baltimore, MD. Savannah, GA: National Coalition for Marine Conservation, Inc.

Quinn, F.H. 2002. Secular changes in Great Lakes water level seasonal cycles. *Journal of Great Lakes Research* 28(3):451–465.

Rodriguez, K.M. and R.A. Reid. 2001. Biodiversity investment areas: Rating the potential for protecting and restoring the Great Lakes ecosystem. *Ecological Restoration* 19(3):135–144.

Saxon, E.D. 2003. Adapting ecoregional plans to anticipate the impact of climate change. In *Drafting a Conservation Blueprint: A Practitioner's Guide to Planning for Biodiversity*, ed. by C.R. Groves, 345–365. Washington, DC: The Nature Conservancy, Island Press.

Shuter, B.J., and K.K. Ing. 1997. Factors affecting the production of zooplankton in lakes. *Canadian Journal of Fisheries and Aquatic Sciences* 54:359–377.

Shuter, B.J., and J.D. Meisner. 1992. Tools for assessing the impact of climate change on freshwater fish populations. *GeoJournal* 28:7–20.

Sly, P.G., and W.D.N. Busch. 1992. Introduction to the process, procedure, and concepts used in the development of an aquatic habitat classification system for lakes. In *The Development of an Aquatic Habitat Classification System for Lakes*, ed. by W.D.N. Busch and P.G. Sly, 1–13. Boca Raton, FL: CRC Press.

Solomon, S., D. Qin, M. Manning et al., eds. 2007. *Contribution of Working Group I to the Fourth Assessment Report of the Intergovernmental Panel on Climate Change*. Cambridge and New York: Cambridge University Press.

Sousounis, P.J., and E.K. Grover. 2002. Potential future weather patterns over the Great Lakes region. *Journal of Great Lakes Research* 28(4):496–520.

Wilcox, D.A. 2004. Implications of hydrologic variability on the succession of plants in Great Lakes wetlands. *Aquatic Ecosystem Health & Management* 7(2):223–231.

Wilcox, D.A., J.E. Meeker, P.L. Hudson, B.J. Armitage, M.G. Black, and D.G. Uzarski. 2002. Hydrologic variability and the application of Index of Biotic Integrity metrics to wetlands: A Great Lakes evaluation. *Wetlands* 22(3):588–615.

Wuebbles, D.J., and K. Hayhoe. 2004. Climate change projections for the United States Midwest. *Mitigation and Adaptation Strategies for Global Change* 9:335–363.

Yediler, A., and J. Jacobs. 1995. Synergistic effects of temperature-oxygen and water-flow on the accumulation and tissue distribution of mercury in carp (*Cyprinus carpio* L.). *Chemosphere* 31:437–453.

Climate Change and Biodiversity in the Great Lakes Region

From "Fingerprints" of Change to Helping Safeguard Species

KIMBERLY R. HALL AND TERRY L. ROOT

OVER THE LAST CENTURY, THE AVERAGE GLOBAL SURFACE TEMPERA-
ture has increased approximately 0.8°C, and the rate of warming continues
to accelerate (Trenberth et al. 2007). Even with this amount of warming,
which is small compared to the net increase we may see in the relatively near
future (an additional 1.1° to 6.4°C or more increase in the global average
by 2100 according to Meehl et al. 2007), wild species are already exhibiting
discernible changes (Root et al. 2003; Parmesan and Yohe 2003; Parmesan
2006). Like other regions at moderate latitudes, temperature change projections for the Great Lakes region are somewhat higher than projections for
the global average. Temperatures in winter are rising faster than any other
season, which may seem like a welcome change to some residents, but can
lead to costly impacts in agricultural and managed forest systems through
increases in the survival of crop and forest pests (CCSP 2009; Bradshaw
et al., this volume). Specifically, winter temperatures will be "less cold"
(nightly low temperatures are expected to increase more than the daytime
highs), contributing to lengthening of the frost-free growing season, which
has already increased by more than one week (Field et al. 2007; CCSP 2009;
Andresen, this volume). By the end of this century, average summer temperatures are projected to increase by 3° to 7°C, leading to dramatic increases
in the frequency of heat waves, especially if the temperature increase is at

the higher end of this range (Christensen et al. 2007; CCSP 2009; Hayhoe et al. 2010b). Summer surface water temperatures in the upper Great Lakes (Michigan, Huron, Superior) are currently increasing even faster than the air temperatures, and these changes are triggering a whole range of system-wide impacts, including increases in wind and current speeds, and increases in the duration of the stratified period (Austin and Coleman 2007, 2008; Desai et al. 2009; Dobiesz and Lester 2009).

The rate at which these temperature changes are occurring suggests that many, if not most, wild species will experience climate change as a stressor that reduces survival and/or reproduction, and thus has strong potential to lead to population declines, or even extinction. The most recent Intergovernmental Panel on Climate Change (IPCC) report, which represents the consensus view of a team of hundreds of scientists from across the globe, suggests that 15 to 40 percent of known species will be at an increasingly high risk of extinction as global mean temperatures reach 2° to 3°C above pre-industrial (or 1.2 to 2.2°C above current) levels (Field et al. 2007, based on work in Thomas et al. 2004). Identifying the most vulnerable species, and what we can do to safeguard their populations, is a major challenge, and each day that passes without reductions in the gases that trap heat in our atmosphere (carbon dioxide, methane) stacks the odds further against the survival of sensitive species and systems.

Here we review observed and predicted impacts of climate change on biodiversity in the Great Lakes region, with the goal of helping to inform and motivate actions that reduce emissions, and to promote actions that help safeguard species and systems so that they can adapt to ongoing changes. Although the scientific literature documenting climate change impacts on species is rapidly expanding, predicting how and when focal species will respond is a daunting task, in part because changes in temperature and other factors are happening within a climate system that already exhibits high natural variation across both space and time. The observed changes, along with ecological theory, allow us to develop "rules of thumb" for how species are likely to respond to the most direct aspects of climate change (e.g., changes in air or water temperature). In addition, experimental studies or predictive models may provide clues as to how several climate factors (temperature, precipitation pattern) may interact. However, it is important to recognize from the start that because many climate factors and interacting species are changing simultaneously, species may show very complex responses, and thus it is often very hard to categorize relative risk. Predictions of responses to change and the relative vulnerability of species become even more uncertain when we try to put them in the context of all of the other stressors that wild species and ecosystems currently face—such as habitat loss, invasion by non-native species, changes in hydrology, and pollution (see Mackey, this volume)—and changes they will face in the future, including actions that societies take in response to changes in climate.

Understanding how climate change will impact species is further complicated by the fact that several aspects of climate change involve feedback loops, or have the potential to lead to tipping points (dramatic shifts in a system in response to an incremental change in some climate factor). For example, surface water temperatures of the upper Great Lakes are showing summer temperature increases that exceed regional temperature increases on land, in part due to positive feedbacks on the warming rate due to reductions in ice cover. Specifically, ice reflects energy from the sun and insulates the water from the warming air, but melts more quickly when the air is warmer, and this loss of ice cover accelerates the rate of surface water warming (Austin and Colman 2007, 2008; Dobiesz and Lester 2009). Study of lakes also provides a good example of a critical temperature threshold, as areas that are deep enough undergo summer stratification, which is triggered by warming of the surface layer above about 4ºC (McCormick and Fanenstiel 1999). When bodies of water stratify, they separate into a warmer, oxygen-rich top layer (the epilimnion), a zone of rapid temperature change (the thermocline), and a colder bottom layer (the hypolimnion), where decomposition of dead plant and animal matter is a dominant process, and oxygen can become limiting. To make a complex story even more so, recent research demonstrates that the fact that the lakes are warming faster than the air is also important to the process of stratification in large lakes, as decreases in the air-to-lake temperature gradient leads to increases in surface wind speeds, which increases currents and acts in combination with the temperature to further lengthen the stratified period (Desai et al. 2009).

Why are we concerned about how climate change factors may drive changes in the timing or duration of stratification? To put it simply, changes in temperature, both direct and through the ice and wind-related mechanisms described above, have the potential to profoundly change how large lakes in our region function. The differences in temperature, light availability, and other factors that occur as a result of stratification provide a diversity of habitats within stratified lakes, which allows species with a wide variety of temperature and other habitat requirements to persist. The timing of stratification, as well as the timing of the fall "turnover," when the oxygen-rich surface waters cool and increase in density, and finally sink down and mix with the others, can be a critical factor influencing the viability of lake species, especially cold-water fish. Given that changes in temperatures for the upper Great Lakes are projected to continue to match or exceed the air-temperature increases, we should expect to see longer stratified periods and increased risk of oxygen deficits below the thermocline in late summer (Magnuson et al. 1997; Jones et al. 2006; Dobiesz and Lester 2009). Increases in the duration of the stratified period of over two weeks have already been observed for Lake Superior (Austin and Colman 2008), and projections for the end of this century suggest that we could see lakes stratify for an additional one and a half months (Lake Erie for a lower emissions scenario and thus less climate change) to three months (Lake Superior under the assumption of higher future emissions; Trumpickas et al. 2009). As the depth and latitude of a lake, lake basin, or bay decreases, it is less likely to show stratification, but some shallow-water bodies will exhibit oxygen-poor "dead zones" because shallow water warms more rapidly, and warmer water holds less oxygen and leads to increases in respiration rates for aquatic species. As warming continues, we should expect more and more areas to develop "dead zones," and for others to transition from stratifying in summer to not stratifying at all, with resultant loss of species that depend on habitats characterized by colder water (see Mackey, this volume, for more discussion of the causes and impacts of dead zones).

Many of us recognize that the Great Lakes moderate temperature changes in nearshore areas within our region, and as a result, these huge water bodies may be perceived as being somewhat protected from the impacts of changes in climate. The fact that at least the upper lakes are warming and changing faster than most land areas in our region may be surprising to many, and this fact challenges us to change how we think about these dynamic, critically important ecosystems. We argue that this well-documented "surprise" of rapidly warming large lakes is likely to be one of many, and we must be prepared for surprises that will change ecological conditions such that the knowledge we have relied upon to inform management and conservation practices may no longer be relevant. First and foremost, we need to act to minimize the degree of change, which means we need to effectively and efficiently reduce our emissions. Yet, as is clear from the examples in this and many other chapters in this volume, we are already seeing the impacts of climate change in the Great Lakes region, and so we also need to help people adapt, and safeguard the species and ecosystems upon which we depend. To move forward on adaptation, we will need to anticipate changes, learn from the results of our actions, and be prepared to quickly change course in response to the unexpected. The first steps in taking action to protect species are to review what we already know, and to see how ecological theory can help us understand what changes in species and ecosystems are likely.

Our review of the impacts of climate change on wild species begins with and focuses on responses to increases in air and water temperature. We spend the most time on temperature, rather than other key factors like precipitation patterns, lake-level changes, or direct effects of increased CO_2 concentrations, for three reasons. First, mean annual temperature has shown a strong and consistent pattern of increase, and projections from various Global Circulation Models (GCMs) show strong agreement in their predictions of increasingly rapid temperature rise (Meehl et al. 2007). Second, changes in many species can be statistically attributed directly to the anthropogenic (human-caused) component of this temperature increase (Root et al. 2005; Rosenzweig et al. 2007), and further temperature changes will take place in the context of geographic patterns that are well understood (i.e., air temperatures cool as you go toward the polar regions or up in altitude; in summer, deep water is cooler than surface water). Third, there is a well-developed body of ecological theory that helps us to frame predictions of how species will respond to increases in the various facets of temperature, such as warmer summers, longer growing seasons, and reductions in the number of days with temperatures below freezing. Often, there are many mechanisms through which the same general impact is realized: for example, warmer waters can stress species because the increase in temperature reduces the oxygen-holding capacity of water, and because at higher temperatures, the respiration rate of many species, which determines how much oxygen is needed, is higher.

Quantifying responses of wild species to climate change is a major challenge, as shorter-term (e.g., monthly, yearly) variability in temperature, as well as the wide variety of other factors influencing species, leads to lots of "noise" in our scientific data. Thus, to detect directional patterns within data sets of species and temperature changes, we rely on long-term data sets (e.g., over 20 years) to evaluate and document wild species' responses. Just a decade or two ago, such data sets were relatively rare in the published literature, but as awareness of climate change has grown, data originally collected for a wide variety of different purposes have been linked with temperature for evaluating the responses of species and ecological systems to climate changes (Parmesan 2006). These long-term studies, a small subset of which are described below, help us to understand current stresses on species and current differences across species groups in terms of type and degree of response, and help us to anticipate future responses as the climate continues to change.

OVERVIEW OF OBSERVED CHANGES

Responses already observed in wild animals and plants to warming temperatures can be grouped into five basic types: (1) spatial shifts in ranges boundaries (e.g., moving north in the Northern Hemisphere); (2) spatial shifts in the density of individual animals or plants within various subsections of a species' range; (3) changes in phenology (the timing of events), such as when leaves emerge in spring, or when birds lay their eggs; (4) mismatches in the phenology of interacting species; and (5) changes in genetics. These categories are not mutually exclusive—for example, a change in the timing of bird migration can represent both a phenological shift and a shift in gene frequencies (genetics). Further, it is important to recognize that shifts in density and abundance include extirpation (loss of a species from a local area) and extinction. The categorization depends primarily on the organizational level (populations, individuals, genes) at which the relationship between a species and its environment were studied. As temperatures continue to rise, and our understanding of other types of climate change impacts improve, we can expect that the volume and variety of impacts on wild species attributable to climate change will continue to increase rapidly as well.

A majority of wild species show predictable changes in responses to increasing temperatures, and the role of temperature in shaping species' life histories is strong. In other words, temperature regime is a key element to which species have adapted over long (evolutionary) time periods. The strong role that temperature plays can be observed in similarities between the edges of species ranges and mapped temperature patterns, north-south patterns in body size within the same species (northern animals are typically a bit larger), and geographic variation in seasonal activities like hibernation and timing of breeding (Root 1988; Brown 1995; Millien et al. 2006). The potential effects of temperature changes are most apparent for ectothermic ("cold-blooded") animals such as insects, reptiles, and fish, for which body temperature, the key determinant of their metabolic rate, strongly tracks the environmental temperature. At lower environmental temperatures, disruption in the availability of energy influences a wide array of physiological and behavioral traits, such as activity patterns and rates of growth and reproduction. In a warming Great Lakes region, we expect that ectotherms like snakes will have longer active periods (prior to becoming dormant for the winter), but higher temperatures, especially in the tropics and deserts, may put these species at high risk if they cannot use behavioral methods to keep cool (Kearney et al. 2009). Homeothermic ("warm-blooded") animals—birds and mammals—maintain a relatively constant body temperature, but still can experience heat-related stress as temperatures continue to increase, especially when they inhabit areas where they are already close to thermal tolerance limits.

Before we give examples of the five types of changes we list above, we want to emphasize that not all changes in species characteristics (phenotypes) require a change at the genetic level. Individuals of many species are able to show flexible responses to temperature as conditions vary among years. Thus, when conditions change in a given location, we can expect to see both "flexible" changes in some species (phenotypic plasticity), and heritable changes (evolution—a change in how common given genes are within the population). In general, phenotypic plasticity can be thought of as a "short-term" solution, as the limits to these responses will eventually be exceeded as a population experiences a long-term increase or decrease in an environmental factor (Gienapp et al. 2008). The potential for evolution in response to climate change is constrained by the degree to which genetic variation for particular traits is present in a given population (Holt 1990). For example, traits that contribute to increased heat or drought tolerance must be present in a population for natural selection to favor the individuals that have those traits, and eventually lead to an overall change in the proportion of individuals that have that "adaptive" trait in later generations. For many of the Great Lakes region's species of greatest conservation concern, we already suspect that population declines, habitat fragmentation, and other stressors have reduced the level of genetic variation such that there is little variation left upon which natural selection can act; however, it is exceedingly rare to actually have data on genetics over time that can be used to confirm or refute this suspicion. Similarly, evidence for genetic responses to climate change is extremely rare, as it requires genetic data to have been sampled over time (Balanyá et al. 2006; see the Genetic Change section below for a few examples). Further, for many species, including some that are able to show flexible responses within a limited range of temperature increases, genetic changes are likely to occur too slowly for natural selection to keep pace with the rapid warming in the environment. As species "fall behind" in terms of adapting to changing conditions, we are highly likely to see more examples of reductions in fitness, population declines, and eventual extinctions. In addition, species that are able to adapt quickly to new conditions may put additional pressures (e.g., as competitors, predators, or parasites) on those that lack sufficient genetic variation, further accelerating the process of species loss.

SHIFTS IN RANGE BOUNDARIES AND ABUNDANCE PATTERNS

The first two types of observed changes on our list, shifts in range boundaries and changes in abundance patterns, both result from changes in the same key "vital rates" (i.e., birth and death rates), but differ in terms of whether we are focused on larger scale (species' ranges) or smaller scale (local abundance) impacts. For many species, changes in climate conditions will enhance a given species' survival rate, growth rate, and/or reproductive rate in some parts of the species' range, and reduce one or more of these rates in other locations. Thus, even without dispersal (movement away from previously occupied habitats), these changes can lead to shifts in the subset of areas within a range where species are common, rare, or absent, and eventual changes in range. Changes in vital rates like survival can be linked back to the physiological constraints of balancing energy reserves under specific climatic conditions, as individuals in highly suitable climatic conditions will often have higher reproduction, survival, or both, than individuals in habitats that are more "costly" (e.g., higher cost of foraging due to heat or cold stress, higher metabolic rate due to higher water temperature for aquatic species).

For a long-term shift into new areas to occur, there must be a way for species to move, a path for them to follow, and a place to go that has climatic conditions that will permit individuals to survive and reproduce. As a general rule, range shifts in response to warming temperatures result in species moving to higher latitudes or altitudes. Movements in mobile species can be direct responses to temperature, such as fish seeking out deeper, colder water, or can be the result of natural selection acting on more random movements by populations of individuals, as those that become established in areas with more suitable climates are more likely to survive and reproduce. Similarly, for species like plants, which are rooted in one location, shifts in range occur as a result of a life stage like seeds being dispersed (e.g., by wind or birds) and becoming established in new areas that are now presumably more suitable than they had been in the past. As much of the Great Lakes region is fairly flat, shifting upwards in altitude is typically not an option; thus, it is likely that terrestrial species would have to migrate for long distances (hundreds of kilometers) over relatively short time periods (several decades) to "follow" their preferred temperature regime. As a result, species that can move rapidly (e.g., birds) are typically seen as more likely to be able to keep up with climate change than other species with lower dispersal capacities (e.g., amphibians, most plants), although of course birds and other mobile species depend on many plants and insects for food and shelter. In addition to moving north within river systems or large lakes, as noted above, some aquatic species also may be able to move into deeper, cooler waters within the same water body, although these deeper habitats may not have all of the other resources that a given species requires.

Examples of species showing range and abundance changes in the Great Lakes region are beginning to accumulate, with the best documented examples coming from researchers conducting long-term research on topics such as community composition and population dynamics. Work by Myers and colleagues (2009) on mammals in Michigan documents rapid changes in ranges for several common species, including northern range edge shifts of over 225 kilometers since 1980 for white-footed mice (*Peromyscus leucopus*) and southern flying squirrels (*Glaucomys volans*). The movement of white-footed mice is of concern from a public health perspective, as these mice are key hosts for the ticks that carry Lyme disease (Ostfeld 1997). The work of Myers et al. (2009) also provides an example that many throughout our region may have recognized in our own lifetimes, and we would not even have had to leave our cars to see the evidence of change. Using distribution data from roadkill surveys conducted in two time periods, 1968 (collected by the Michigan Department of Natural Resources and published in Brocke 1970) and 2006–2008, they document a rapid northern range expansion of our region's only marsupial mammal, the common opossum (*Didelphis virginiana*). In figure 1, we have matched the two data sets from Myers et al.'s work with maps of changes in winter (December–February) minimum temperatures, as this aspect of temperature is likely to have an important influence on where opossums can survive. Note that now, lakes Michigan and Huron seem to form a barrier for further movement into the eastern Upper Peninsula, but opossums seem to be shifting into the western Upper Peninsula from Wisconsin. These three examples highlight species moving in, but Myers et al.'s work also documents reduced ranges for more northern species, and suggests that at least at some well-studied sites, species from the south are replacing those that were formerly dominant, rather than adding to species diversity.

Species are also showing changes in abundance within current ranges. Studies on moose (*Alces alces andersoni*) provide an indication of the complexity of the sensitive relationship between a species' population numbers and environmental temperature. Two separate research teams focused on understanding factors such as birth rates, parasite loads, and survival of moose in northwestern Minnesota (Murray et al. 2006), and on Isle Royale (Vucetich and Peterson 2004; Wilmers et al. 2006). They suggest that warming temperatures are contributing to local population declines. The Minnesota team found that the population growth rate over 1961 to

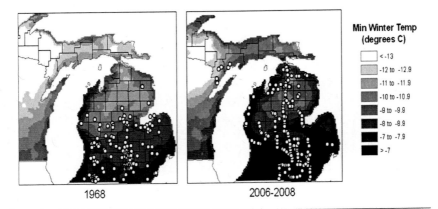

Figure 1. Observations of road-killed common opossum from 1968 and 2006–2008, from Myers et al. (2009, figure 4), overlaid upon 20-year average values of minimum winter (December–February) temperature from the period prior to the survey dates (1949–1968, and 1987–2006). The 1968 opossum data were reported by Michigan Department of Natural Resources personnel and presented in Brocke (1970), and were digitized from this source by Myers et al. (2009). The 2006–2008 data were collected by Myers and colleagues, who note that while much of the Upper Peninsula was surveyed, the southwestern corner and eastern edge of the state were not covered. The temperature data are from the PRISM data set at 4 km resolution (Gibson et al. 2002) as available from ClimateWizard (2009).

2000 was negatively associated with higher temperatures, and both teams suggest that warmer temperatures stress moose in a wide variety of ways, from increasing the rate of heat-stress deaths and the impact of parasite infections, to decreasing feeding and birth rates. Both study locations are in the southern part of the species' range, and modeling by the Minnesota team suggests that, given the observed relationships between vital rates and temperature, the northwest Minnesota population of moose will not persist over the next 50 years (Murray et al. 2006). In related work on the dynamics of host-parasite systems in Arctic populations of caribou and musk ox, Kutz et al. (2005) found that one type of nematode (worm-like) parasite can now complete their life cycles in a single year, rather than the three or four years required for reproduction in colder climates. This change has led to rapid increases in the abundance of one species (the parasite) at the expense of the survival of another (the mammalian host; Kutz et al. 2005). Ecologists that specialize in disease dynamics caution that increases in the impacts of current diseases and pathogens are only one climate-change-linked threat to the Great Lakes region's iconic large mammals: as the ranges of other mammals shift farther north, they will likely carry with them diseases and parasites to which current residents of these habitats have little or no natural resistance (Kutz et al. 2005; Dobson 2009).

Changes in species' range boundaries and abundance patterns are of concern for several reasons. First, the rapid changes in climate are taking place in the context of a wide range of other impacts on natural systems, most notably habitat loss and habitat fragmentation due to conversion of natural areas into farms, cities, ports, and other land uses. Further, we expect the impact of humans on the landscape and shoreline to increase as we shift the location of human-changed areas in response to changes in drought risk or lake level, or the need to use more land to support greener energy sources (e.g., wind and solar energy, production of crops and woody bio-mass for use as biofuels). Highly mobile species like birds and mammals are more likely than many others to be able to shift their ranges fairly rapidly; however, these movements are frequently slowed or blocked by inhospi-table landscapes. In the Great Lakes, barriers for terrestrial species can range from the lakes themselves (recall the lack of opossums in the eastern Upper Peninsula of Michigan in figure 1) to large expanses of agricultural fields or pavement. Similarly, dispersal by aquatic species can be limited by the geography of the water body they inhabit, and by human-made barriers such as dams. Ironically, in the Great Lakes region, some of the aquatic barriers have been put into place to protect aquatic systems from damaging invasive species like sea lamprey (*Petromyzon marinus*), so we cannot simply remove them to help species move to more suitable habitats. Even in areas where we have large expanses of intact ecosystems, increasing temperatures can make habitats that depend on particular levels of surface or groundwater more fragmented. Specifically, as evaporation rates increase due to warmer temperatures, even large expanses of coastal wetlands, inland marshes, or wet prairies can become less connected for species like frogs or ducks as some sections dry out, and the remaining wet areas upon which many spe-cies depend become more isolated, or disappear.

Second, range and abundance changes are of concern because species that are not able to disperse will have the added stress of species from lower latitudes and altitudes invading their habitats. So, individuals at the south-ern end of their species' range have the potential to be stressed both by climatic conditions that are becoming less and less favorable, and by species that move in from warmer areas and are less challenged by the same climatic factors. The species moving in may directly compete for key resources, and also may contribute to the decline of resident species by spreading dis-eases and parasites. In the Minnesota moose example described above, the authors suggest that temperature impacts on moose are increased in areas where a more southern species, white-tailed deer (*Odocoileus virginianus*), are abundant, as the deer act as a reservoir for parasites (Murray et al. 2006). In addition to species that act as stressors on wild species of conservation concern, range shifts by species that act as forest or crop pests, or that are detrimental to public health (i.e., carry diseases, create toxic algal blooms) are key concerns in the Great Lakes region, and are important subjects of observational and model-based research studies.

Third, we are concerned about range and abundance shifts because species movements will often be independent of shifts of other species. We expect species to shift independently, as the set of constraints that describe the habitat and ecological niche for each species (factors like temperature, food availability, soil types, and stream flow characteristics) is unique. In effect, we expect to see the "tearing apart" of sets of species that typically interact, and many of these interactions may be critical to the survival of one or more of the interacting species. We do have some examples of "specialist" butterfly species increasing the variety of habitats that they use as they shift north (Thomas et al. 2001), but in general, we consider strong dependence on one or a few species to be a strong indicator of vulnerability to climate change. To give an example from the Great Lakes region, the federally endangered Kirtland's warbler (*Setophaga kirtlandii*), which breeds almost exclusively in northern Michigan, is at risk due to a strong dependence on young jack pine trees (*Pinus banksiana*). Jack pine trees grow on sandy, nutrient-poor soils, and are tolerant of cold spring temperatures that can often damage or kill other tree species. For the past three decades, the warbler has shown a dramatic population increase as more and more habitat has been made available, primarily as the result of an intensive collaborative effort between state, federal, and private landowners in Michigan to create large expanses of young jack pine (Donner et al. 2008). However, maintaining jack pine habitat is likely to get more and more challenging as a result of increased drought stress in the summer, increased competition from other trees that are favored by the warmer winters and springs, and increases in the abundance of insect pests formerly limited by cold weather. If our goal is to keep providing habitat for the warbler over the long term, we need to be thinking about how to help facilitate a shift in range, given that the proper soils are relatively rare, and about how to modify management strategies in places that cannot be replaced.

CHANGES IN PHENOLOGY

In many organisms, seasonal changes in temperature act as cues that trigger transitions in the species' seasonal cycle, such as metamorphosis (e.g., the transition from egg to larvae), or development of new leaves. The dominant cue for some seasonal changes, like the start date for migration for many birds, is a change in day length (Berthold 1996), but temperature can still have a strong influence on the timing of migration by influencing the rate at which birds travel from their wintering grounds to breeding habitats. In addition to directly triggering changes in timing, warming trends can impact species by influencing other key seasonal events that trigger changes in their seasonal cycles, such as timing of snowmelt or flooding, or lake stratification. The term "phenology" describes the timing of these seasonal events, and recording the date of events like bird arrivals, tree fruiting, or ice breakup on lakes has been of interest to both scientists and members of the general public for decades—and even centuries in some locations. As a result of this broad interest in the timing of seasonal events, the scientific community has had access to long-term data sets showing strong, easy-to-observe responses by species to changes in temperature. These long-term data sets have played a key role in the evaluation of climate change impacts, and in raising public awareness about the risks to biodiversity posed by what many might think of as "inconsequential" changes in temperature.

The strength of phenological data as a tool for raising awareness of the impacts of climate change is hard to overemphasize. When aggregated across large scales, these data show very strong patterns, and because many people are familiar with the types of events (e.g., timing of bird arrivals or leaf opening in spring) that are being measured, changes can be clearly communicated to nonscientists. Assessing broad patterns in responses to climate change is possible with phenology data because various types of changes seen in a wide range of species can be measured in the same unit: number of days earlier or later. This ability to group observations of change across species and continents using the same measure of change has facilitated broad "meta-analyses" that link shifts in timing to local changes in temperature (e.g., Root et al. 2003, 2005; Parmesan and Yohe 2003), and has facilitated a rapid increase in the recognition of climate change as a threat to biodiversity. Several early phenology studies that were highly influential in raising awareness that species were responding to changes in climate focused on, or included, study sites in the Great Lakes region. These included evidence of 10- to 13-day advances in frog-calling dates (an indicator of breeding) in western New York in response to a 1° to 2.3°C increase in temperature in key months (Gibbs and Breish 2001), advances in the timing of many spring events (bird arrivals, plant blooming) on a Wisconsin farm in the 1980s and 1990s relative to observations taken by Aldo Leopold in the 1930s and 1940s (Bradley et al. 1999), and a nine-day advance in the laying date of tree swallows (*Tachycineta bicolor*) across the continental United States over 32 years (1959–1991; Dunn and Winkler 1999).

Overall, the patterns of change documented in phenological studies provide strong evidence for a "fingerprint" of climate change on the timing of seasonal events. In the simplest form of comparison, phenological responses to increasing temperature can be grouped into three patterns: shifts toward earlier phenology, shifts later, and no change (Root et al. 2003; Parmesan and Yohe 2003; Root et al. 2005; Parmesan 2006). Focusing on the first two groups (shifts earlier and later), we would expect to see a roughly equal (or random) number of species exhibiting each type of change if increasing temperature was not a causal factor. Instead, in a comparison by Root et al. (2003) of published records of timing of spring events from 700 species exhibiting statistically significant change over the last 30 years in locations from across the globe, only 6 out of the 700 species (<1%) showed a shift toward later timing. Overall, these data suggest a strong pattern of spring events occurring earlier across different types of species and around the world, at a rate of around 5 days per decade (Root et al. 2003).

In addition to helping us understand the direction and rate of change that species are exhibiting, phenological data have also played an important role in linking shifts in species to human actions that contribute to increasing CO_2 like burning fossil fuels and deforestation. Root et al. (2005) compared temperature predictions from global circulation models (GCMs) derived from model runs using only natural forcings (e.g., volcanic dust), only human forcings (e.g., CO_2 emissions), or a combination of both natural and human forcings. By comparing the observed changes in species with these differently modeled temperature predictions, they were able to see the influence of natural and anthropogenic drivers of change. When only natural forcings were used, the associations between the predicted temperatures and observed changes in species were quite weak, while with only anthropogenic forcings, the correlations were stronger, and with a combination of both forcings, the fit between modeled and observed changes was quite strong (Root et al. 2005).

Quantifying the overall percentage of species that are changing is harder than comparing the proportions showing shifts earlier or later. This is because we know that some species are not exhibiting changes in the timing of their spring events, even in locations where temperatures have increased over time. However, when we try to capture global trends in a review of literature published over many decades, we run into the problem of publication bias. Specifically, it is much less likely that research documenting "no change," or an insignificant relationship between temperature and a change in timing, will be published unless this information is reported along with reports of significant change in another species. There are several reasons why changes may not be detected; some changes may be triggered by other factors like day length, or seasonal weather may be highly variable, such that it takes several decades or longer than the duration of a given study to detect a statistically significant trend. However, lack of a phenological response does not mean that species are "immune" to temperature changes, for at least two reasons. First, they may be responding in other ways not addressed in the same study, for example by changing behavior or distribution, and second, a lack of response may have negative consequences (see "Phenology Mismatches" discussion, below). Multi-species monitoring networks provide essential data sets for helping us tease apart what factors determine which species respond, and the consequences of responding or not. Thus, continued support of new and existing networks of monitoring is an important investment that will help us determine what actions to take to better protect species over the long term.

Although examples abound of changes in phenology, indications of what these changes mean in terms of risk to species from climate change requires comprehensive data sets that go beyond just measures of timing. A recent paper documenting long-term (approximately 100 years) changes in phenology and abundance of 429 plant species in Concord, Massachusetts (many of which are also found in the Great Lakes region) showed that although there has been an overall shift of 7 days in flowering phenology associated with a 2.4°C temperature increase in the study area, some plant families are showing less of a response to temperature than others (Willis et al. 2008). In many cases, this failure to shift flowering time in response to changes in seasonal temperature was associated with strong declines in abundance (Willis et al. 2008). Thus, rather than just documenting a change in timing, this paper represents a major contribution to helping us understand why we should be concerned about phenological changes (or lack of change), as it links the timing change to a change in fitness through the measure of abundance. Of particular interest, Willis and colleagues (2008) found that the plant species showing declines were more closely related than expected by chance. As research continues on both genetic and flexible (phenotypic) responses to climate change, it seems likely that broad patterns will continue to be identified that help us to evaluate the responses of species by group, rather than trying to plan for the idiosyncratic responses of each species.

PHENOLOGY MISMATCHES

Rapid phenological changes are of concern because for tens of thousands of years or more, wild species have been adjusting to seasonal changes in their environment, and to associated changes in species that act as predators, prey, parasites, and competitors. By contributing to the buildup of greenhouse gases, we have dramatically increased the rate of change, and more and more studies are suggesting that species in the same area are not responding at the same rate. Our fourth category of observed changes, phenology mismatches, describes situations where species that interact in some important way respond differently to a temperature change. The potential importance of mismatches may be easiest to imagine in systems where attainment of a threshold temperature cues the emergence of leaves of a dominant tree or grass, or algal growth. In such a system, a shift in the timing of spring warming that alters when these plants grow or bloom could represent a key change in the foundation of the food web that determines energy flows throughout that entire ecological system. If other species in the same system do not shift in the same direction and at a similar rate, they may be at a strong disadvantage in terms of their ability to survive and reproduce relative to other species with similar resource requirements. Evidence of the potential for cascading phenology mismatches in lake food webs comes from work by Winder and Schindler (2004) in Washington State. In their focal lake, they found that phytoplankton (algae) have been blooming earlier, with some types of zooplankton (tiny animals that eat algae) showing similar changes in timing, and others not, potentially impacting food availability for a wide range of fish and other predators (Winder and Schindler 2004). As noted above, conditions in the Great Lakes are changing rapidly (increasing temperature, longer stratified period, stronger currents), suggesting a high potential for species to respond at different rates and contribute to disruption of entire food webs.

Although a wide variety of species are likely vulnerable to phenological mismatches, we are not aware of any research that has focused on this topic within the Great Lakes region. However, it is not very hard to pull together information that makes a compelling argument that these types of changes should be of concern, and we do this here using the example of songbirds that migrate through, and breed in, the Great Lakes region. The northern Great Lakes region stands out within North America for supporting a high diversity of breeding songbird species (Price et al. 1995), and the region also supports vast numbers of birds during spring and fall migration. Most songbirds depend upon a ready source of insect prey, both along their migration routes and in their breeding habitats. Studies in Europe have documented advances in insect emergence relative to bird arrivals at breeding habitats, and suggest that these timing mismatches are leading to reduced breeding success (Both and Visser 2001; Visser et al. 2006).

To assess the vulnerability of migratory birds to changes in plant or insect phenology, we need a "yardstick" for understanding the impact of a particular magnitude of change (or lack of change) in phenology (Visser and Both 2005; Visser 2008; Both et al. 2009). The work cited above, along with work by Marra et al. (2005), suggests that changes in bird migration phenology may be slower than the responses of many of the plants and insects at the stopover sites upon which these birds depend. Marra et al. (2005) compared the median capture dates of 15 long distance migrants from bird monitoring stations in coastal Louisiana and two stations in the Great Lakes region, Long Point Bird Observatory (on the north shore of Lake Erie) and Powdermill (western Pennsylvania). They also compared the duration of time between the median arrivals for the same species at the southern and northern sites. Marra et al. (2005) found that median capture dates were earlier in years with warmer spring temperatures (mean April/ May temperature) for almost all of their focal species, at a rate of roughly 1 day earlier per each 1°C increase in temperature. However, they note that lilac (*Syringa vulgaris*) budburst occurred 3 days earlier for the same temperature increment, a similar rate to the average reported for plants in the Willis et al. (2008) study described above. Similarly, Strode (2003) suggests that North American wood warblers are not advancing in phenology as fast as key prey are likely to be responding to increased temperatures (e.g., the eastern spruce budworm, *Choristoneura fumiferana*). Marra et al.'s results of earlier median passage in warm springs is still consistent with our earlier statement that a key cue for migration departure is day length; earlier arrivals were at least in part achieved through faster migration (as opposed to earlier departure dates) as the duration of migration between the southern and northern locations decreased by 0.8 days with every 1°C increase (average of 22 days; Marra et al. 2005).

In the Great Lakes region, migratory songbirds may be most at risk by being at the right place at the wrong time at migratory "stopover" habitats at the edges of the upper Great Lakes, due to rapid warming of lake

surface waters relative to nearby land (described above, Austin and Colman 2007, 2008; Dobiesz and Lester 2009). This pattern suggests that even
if birds manage to track the rate of phenological shifts occurring on land
near the shoreline, they may be too late to take advantage of the emergence
of aquatic insects upon which many birds depend while refueling during
migration (Ewert and Hamas 1995; Smith et al. 1998). If this occurs, the
timing mismatch could cascade through the migratory process, with birds
having to stay longer at the stopover site in order to regain energy reserves
for sustained flying (Moore et al. 2005).

Even without considering whether coastal stopover sites are warming
faster than inland sites, changes in phenology along migration routes are
likely to increase the risk of mortality, or at least the potential for reduced
fitness, of migrating birds. This is because migratory birds are likely to
encounter strong geographic variation in the extent to which temperature
trends, and as a result plant and insect phenology, are changing: Some sites
show strong warming trends, while others may have warmed less over time,
or even cooled. To illustrate this point, in figure 2 we have drawn two hypothetical migration routes for birds breeding in the northern Great Lakes
region, and show them over a map of trends in spring maximum temperature for a 50-year period (1957 to 2006). Note that although the central
United States is experiencing an overall annual warming trend, maximum
spring temperature, a factor that is likely highly correlated with plant and
insect phenology, has been showing both increases (dark areas) in some
areas, and even slight decreases (white to light gray) in other areas during this time period. So if birds were to follow the arrows in the diagram,
though they would encounter a lot of local variation, in general they would
first encounter areas that are showing moderate warming (e.g., Texas, Florida), then reach areas that are warming less or even cooling (Arkansas, much
of Tennessee), and finally reach areas, potentially including their breeding
grounds, that are warming much more rapidly (much of Minnesota, western
Wisconsin, and Michigan). Thus, while species that migrate are obviously
highly mobile and able to rapidly shift their distribution, the potential for
reduced synchrony with key resources along migration routes suggests that
they may be particularly vulnerable to the impacts of climate change.

CHANGES IN GENETICS

Most studies documenting responses to climate change focus on readily
observable characteristics such as phenological shifts; however, increasing

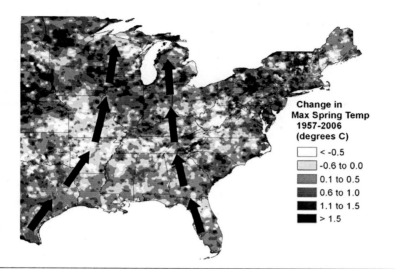

Figure 2. Variation in estimates for net change in the average spring (March–May) maximum temperature over a 50-year period (1957–2006). The arrows indicate two hypothetical pathways that birds might follow as they migrate through the southern U.S. toward breeding grounds in the northern Great Lakes region. Note that although much of our region is showing peak temperature increases, birds will have to move through regions where peak spring temperatures are showing decreasing trends, suggesting birds may encounter high variation in phenology of key plants and insect prey along their migratory routes. The temperature data are from the PRISM data set at 4 km resolution (Gibson et al. 2002), with 50-year trend calculated by ClimateWizard (2009; see Girvetz et al. 2009 for details).

numbers of studies are showing that changes in other characteristics, such as morphology (body shape or size), behavior, and underlying gene frequencies, can be linked to rapidly warming temperatures. Demonstrating changes in gene frequencies is a major challenge, as it requires these frequencies to have been measured for many generations; as a result, most examples are studies of short-lived insects like fruit flies (*Drosophila* species). Work on fruit flies around the world has demonstrated shifts in how chromosomes are arranged that correlate with geographic patterns, e.g., populations in the north shift toward showing patterns like those to the south as climate warms (Levitan 2003; Balanyá et al. 2006; Etges and Levitan 2008). These changes tend to be discussed in terms of "heat tolerance," yet the actual benefit of these changes in terms of enhanced viability have not yet been established (Gienapp et al. 2008).

In another insect example, Bradshaw and others (2000) found evidence that a mosquito in eastern North America is responding to changing growing-season length, even though the trigger for dormancy (the focal phenological event in the study) is related to day length, not temperature. They found that the day-length cue that initiates dormancy in pitcher plant mosquitoes (*Wyeomyia smithii*) is genetically controlled, and warmer fall temperatures have apparently led to selection for individuals that delay the dormancy period, favoring a shift in the day-length trigger toward a shorter day length. Strong evidence of similar genetic changes in vertebrates in response to climate change is very rare (Gienapp et al. 2008), but one notable exception comes from long-term research focused on red squirrels (*Tamiasciurus hudsonicus*) in western Canada. Work by Réale et al. (2003) demonstrated that shifts toward earlier breeding phenology in response to climate-induced changes in food supply are the result of both phenotypic plasticity (87 percent of the change) and an evolutionary response (13 percent).

Although results suggest that some species may be able to respond quickly to changes, many others may lack the genetic variation that might allow selection, and thus adaptation, to occur. In other cases, as has been demonstrated for a Minnesota population of a native prairie plant (*Chamaecrista fasciculata*), adaptive responses can be slowed even when variation is present, due to linkages between traits that are "antagonistic," such that one confers benefits in a new climate, and another does not (Etterson and Shaw 2001). As scientists focus on actions needed to conserve biodiversity, our key concern is that climatic changes are occurring so fast (and are projected to keep increasing in rate) that the genetic changes needed for the survival of a species may not happen quickly enough to keep up with the rapidity at which the planet is warming. Minimizing the rate of future changes in climate is thus an essential part of protecting the Great Lakes region's biodiversity over the long term.

CHANGES IN PRECIPITATION

While our review focuses on the biological theory and observations related to species responses to changes in air or water temperature, many other aspects of species' environments are also changing due to global warming. These include factors like precipitation patterns, storm intensity, drought stress, ice cover, and lake level, all of which may interact with, or even counteract, some effects of temperature increase. When compared to temperature increases, changes in the patterns of precipitation are harder to generalize into "rules of thumb" that can help us anticipate species' vulnerability. For average precipitation in particular, predictions from different GCMs (there are about 16 in common use, as well as many regional models) for the Great Lakes region are highly variable across space and time, and there is more uncertainty in both the direction of magnitude of changes projected, and the time of year in which any projected changes will likely occur. Thus, when thinking about precipitation in particular, probably the most important rule is that we must be prepared to be more adaptive in our conservation and management actions, and to inform these actions using the range of possible outcomes suggested by modeling work. In other words, we need to think both about how species may respond to more rain in various seasons, and to less, and to remember also that even if the amount of rain or snow stays about the same, it will evaporate or melt faster in a warmer climate.

The same "bet hedging" approach should be applied to planning for changes in water levels of the Great Lakes, as these are influenced by air temperature regime (which drives losses of water through evaporation), water temperature regime (including the feedback between ice cover and water temperature described above), and by precipitation (for more detail on lake-level processes, see Mackey, this volume). While projections for future lake levels that focus on mean values suggest steady declines in lake levels over time (e.g., Hayhoe et al. 2010a, also cited in CCSP 2009), work by Angel and Kunkel (2010) demonstrates that when you explicitly consider the variation across GCMs and across emissions scenarios, lake-level models suggest a wide range of potential changes, including slight increases. This variation primarily results from differences across GCMs in projections for the amount and timing of precipitation, which is a persistent challenge for climate modelers, and one that is not likely to be resolved soon. Thus, the most pragmatic approach is to develop conservation plans that anticipate drops in mean lake levels, but that also benefits species if the seasonal range of lake levels stays about the same or even increases. Further, we need to also be thinking about how human use of these precious water resources might change, and how to include protection of ecosystems and species into decisions that are made regarding future water-use regulation policies.

With respect to extreme precipitation events rather than mean values, however, there is general agreement that the frequency of extreme rain events (intense storms) will increase, especially in the winter and spring. Trends over the last 50 years for the upper Midwest suggest about a 30 percent increase in the amount of rain that falls in the top 1 percent of "very heavy" precipitation events, and this impact is expected to increase due to the fact that warmer air can hold more water (CCSP 2009, based on updates to Groisman et al. 2004). The impacts of this change on aquatic systems can be quite strong, especially in landscapes with high proportions of agricultural or urban land uses, which act as sources of pollutants and fertilizers when large volumes of water flow across them into rivers and coastal areas. Similarly, we expect increases at the other end of the extreme weather events spectrum, in the form of summer droughts.

While precipitation changes are hard to predict, there are many things that we can do to safeguard the species and natural systems of the Great Lakes through these ongoing changes. We have this opportunity to do a better job protecting species because many of our land-use and infrastructure decisions directly influence how water moves through our landscapes, and many of the key stressors on biodiversity, especially in aquatic systems, result from impacts that are highly influenced by extreme rain events. As noted above, much of the sediment and pollution that enters rivers and nearshore systems does so during extreme storm events that cause large volumes of water to flow across agricultural fields and cities, and contributes to overflows of combined sewage and stormwater handling facilities. Actions such as restoring wetlands that can absorb water, promoting "best management practices" on farms, and improving our sewage handling and other pollution-prevention infrastructure can all benefit wild species by minimizing pollution and sedimentation from storm events. Further, when actions like wetland restoration, and restoration of functional floodplains, are implemented to help handle stormwater, species also reap the benefit of increased habitat area and enhanced connectivity between habitat patches. These kinds of "green infrastructure" investments also benefit people by reducing the risk of flooding, and by helping to promote clean water for drinking and recreation, and in many cases may provide lower-cost, more reliable strategies for containing flow than traditional infrastructure-based options.

UNDERSTANDING AND RESPONDING TO SPECIES VULNERABILITIES

Thus far, the weight of evidence suggests that the most appropriate expectation for how species may respond to climate change is to anticipate more of the types of changes we have already seen—i.e., changes in ranges (evading the temperature change), and changes in phenology and behavior that allow species to persist in the same range. We need to remember, however, that many of these "stay in place" changes may be the result of phenotypic plasticity, suggesting that in the coming decades, many species that appear to be adapting at a rate that allows them to track changes in climate may show sudden declines in viability once the temperature shift exceeds some critical threshold beyond which their "flexible" response is not enough. As of yet, while there are many examples of changes in species in response to climate change, there are no documented examples of genetic shifts in thermal tolerances that appear to allow species to remain viable in the same location following a change that would have otherwise led to reduced survival or reproduction (Parmesan 2006; Bradshaw and Holzapfel 2008; Gienapp et al. 2008). Even given these caveats, developing our understanding of how we could modify our management and conservation priorities to account for shifts in ranges (i.e., promoting connectivity between habitats) and in phenology (ensuring protection of species across a range of current and future microclimates) is an important place to start making our ongoing work "climate smart."

Based on the wide range of observed impacts described here, it is easy to see that predicting impacts on, and responses by, a given species or system can get complicated very quickly. To illustrate this complexity, we have compiled a list of climate factors that are changing in the Great Lakes region, and examples of how these can impact species (figure 3). Many of these impacts, such as changes in temperature, have broad-reaching impacts, while others are likely to impact relatively small subsets of species through very specific mechanisms. For example, the figure notes that increases in CO_2, the major driver of changes in temperature, can also influence competitive relationships between species. This impact applies specifically to plants, and acts by changing the relative efficiency of different variations on the process of photosynthesis, the mechanism by which plants convert energy from the sun into new plant tissues. Because the suite of potential impacts is so large, and impacts are often interrelated, our best guesses on impacts and species vulnerability may vary considerably depending on how many of these factors are considered. For example, Jones et al. (2006) found that projections of the potential impact of climate change on Lake Erie walleye (*Sander vitreum*) based simply on water-temperature change were very different from results incorporating changes in climate-sensitive factors such as water levels and light penetration. This work relied upon decades of research on this fish's habitat needs and biology, and illustrates that for well-known species like walleye, the challenge to managers and conservation practitioners may focus on characterizing a complex set of direct and indirect climate-related changes that may interact and influence species survival. For most other species, a lack of baseline information from which to even begin the process of understanding potential impacts is often the most daunting challenge.

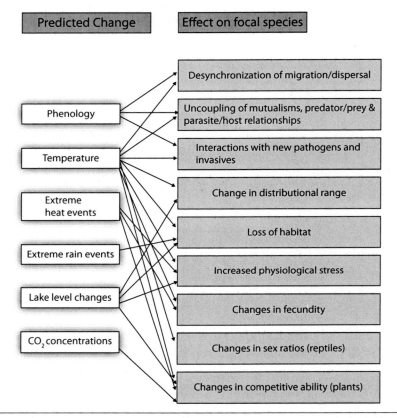

Figure 3. Illustration of the many ways in which climate change can impact species in the Great Lakes region, adapted from Foden et al. (2008, fig. 1). Note that the climate factors and effects are not mutually exclusive—as, for example, the driver "phenology" represents a change in the timing of some factor like temperature or precipitation.

Finally, we will conclude this chapter with some general rules of thumb for identifying highly vulnerable species, and examples of the kinds of "surprises" that will require us to be much more agile in how we plan and implement actions to benefit wild species. Characteristics often identified as indicators of species that are at greatest risk of population decline, or possibly even extinction due to climate change impacts, include:

- Occurrence at high altitude or latitude (cannot shift range farther up or north)
- Occurrence in isolated habitats surrounded by inhospitable land, either developed or undeveloped, that inhibits dispersal

- Being near limits of physiological tolerance
- Having limited dispersal and/or colonizing ability
- Having very specific habitat requirements
- Being highly dependent on interactions with one or a few other species (susceptible to phenology mismatches, and mismatches in rate or location of range shifts)
- Having long generation time (slow potential pace of microevolution)
- Having low genetic variability and/or low phenotypic plasticity.

Although applying these and other much more comprehensive and site-specific approaches to assessing vulnerability is a major challenge, an even greater challenge is likely to be deciding on what percentage of our limited resources should go towards trying to protect those species identified as most vulnerable. Given the projections for rapid increases in species extinctions, we need to think carefully about how to make decisions that protect the widest range of species and functional ecosystems, some of which the public may tend to value, and others that may be less well known or appreciated, but are still critical to the long-term maintenance of biodiversity.

As we work to update our conservation plans and make them "climate smart," it is vitally important that we also update our approaches to management such that they become more agile and able to shift strategies quickly in the face of new information and surprises. Acting in a climate-smart way will also require that we improve our ability to share and synthesize the information we do have, and improve our tools for acting in the face of uncertainty, as discussed in detail in the chapters by Scheraga, Moser, Marx and Weber, and Easterling et al. in this volume. With respect to anticipating surprises, we expect that surprises for resource managers will take at least three forms: (1) exceedance of thresholds (e.g., thermal tolerance thresholds, leading to strong declines in fitness); (2) new interactions among species, and/or new or synergistic impacts related to interactions with climate and other stressors (e.g., invasive species); and (3) higher frequency of extreme weather events with catastrophic impacts on focal systems (floods, ice storms, extreme cold periods in spring).

In some cases, surprises may result from a combination of factors and threshold exceedance, and be surprising due to a mix of positive and negative impacts on fitness. For example, work by Tucker et al. (2008) shows that many red-eared sliders (aquatic turtles, *Trachemys scripta elegans*) in Illinois have been laying an additional clutch (three instead of two) of eggs. The increase in reproduction has been achieved by turtles extending their breeding season 1.2 days per year from 1995 to 2006, and initiating breeding earlier at a rate of 2.2 days per year over the same time period, which showed significant spring to summer warming. However, like other turtles, the sex ratio for this species is temperature dependent, and these later clutches are being deposited into below-ground nests when soil temperatures are low relative to a temperature threshold that promotes an even ratio, leading to a male bias in the offspring. This male bias is surprising, as the general prediction based on warming would be for these turtles to produce more females. However, with turtles nesting earlier and later (first and third clutches), the authors suggest that only the second clutches would be likely to produce many females. Clearly, if the population becomes strongly male-biased, this would suggest population declines.

Some serious risks to public health and wildlife habitat may be in store for us as a result of one species of invasive cyanobacterium, or blue-green algae, that embodies two of the above forms of surprises: thresholds and new invasive species. Recent work by Hong et al. (2006) documents that *Cylindrospermopsis raciborskii*, a toxin-producing and bloom-forming species, has been found in Muskegon and Mona lakes in western Michigan, which connect directly to Lake Michigan. First, *C. raciborskii* is an example of a species that seems to require exceedance of a temperature threshold of roughly 22°C (Hong et al. 2006, and work reviewed therein) to germinate from an inactive, climate-resistant spore, to the motile form that creates blooms and produces toxins in the water column. Second, it has "surprise" potential because this invasive species is a good competitor with other phytoplankton, and we don't know how it moves, or where the spore is likely to be waiting in lake sediments for suitable temperatures to arise (Hong et al. 2006).

In our last example of surprises, we focus on a change that many people who do not depend on income from the Great Lakes region's winter-recreation sector might be looking forward to—earlier spring warming. As described above, we have lots of evidence that species respond to changes in the timing of seasonal events like spring warming, which may suggest that timing changes are not likely to act as stressors on populations. However, in addition to potentially contributing to timing mismatches, shifts in timing may lead to surprises by making species more vulnerable to extreme weather events that are typical for a given season. For example, a period of record low temperatures occurred across much of the eastern United States in April of 2007 after what had been an unusually warm later winter/early spring (Gu et al. 2008). The warm spell led to early bud break and canopy leaf-out in trees, and these sensitive plant tissues were then blasted by a very cold arctic air mass in early April, leading to massive damage to plants across a variety of systems, especially in centrally located states like Tennessee and North Carolina. These kinds of events may be a potential threat to species' viability, both because they can lead to direct reductions in survival or fitness when the event occurs, and because they may have longer-term impacts on future generations by producing selective pressures in opposition to pressures presented by gradual climatic changes (i.e., increases in mean temperature). For example, an early spring cold snap in Nebraska killed a large proportion of cliff swallows (*Pterochelidon pyrrhonata*) that had arrived early following migration (Brown and Brown 2000). In later years, birds at these same colonies arrived significantly later, providing evidence of directional selection on arrival timing (Brown and Brown 2000). Assuming that local and regional warming would favor birds that arrive earlier in most years, intense selection from these rare weather events could be counteracting an otherwise beneficial response.

In conclusion, we would like to reemphasize that based on species responses to climate change that we can already observe, and on the theories that frame our understanding of these changes, models project high rates of extinction if we continue to rely upon carbon-based fuels. Estimates of range and abundance changes in response to various future scenarios (e.g., a doubling of atmospheric CO_2) suggest that between 15 percent and 40 percent (Field et al. 2007, based on Thomas et al. 2004) of all known species will go extinct due to human enhancement of atmospheric greenhouse gases. Given that there are around 1.75 million species that have been described, somewhere between 250,000 and 700,000 known species, to say nothing about the unknown species, could go extinct primarily due to our use of fossil fuels and the dumping of their combustion products into the atmosphere as if it were an unpriced sewer. As a northern region with many rare species, the Great Lakes region is at risk, and we need to take action now to slow the rate of change and protect the biodiversity upon which we as a society depend. Further, we need to act now to help safeguard species against changes that are already occurring, and which we know are coming due to emissions that have already been released. Protecting species and functional ecosystems will require rethinking our management and conservation strategies to ensure that they are climate smart, and that we become much more efficient and targeted in our investments in natural resources. These investments must go beyond on-the-ground actions to include significant investment in tools for managing and sharing information across the region, and tools to help us improve our ability to make decisions in the face of uncertain information.

REFERENCES

Angel, J.R., and K.E. Kunkel. 2010 The response of Great Lakes water levels to future climate scenarios with an emphasis on Lake Michigan-Huron. *Journal of Great Lakes Research* 36 (Suppl. 2): 51–58.

Austin, J.A., and S.M. Colman. 2007. Lake Superior summer water temperatures are increasing more rapidly than regional air temperatures: A positive ice-albedo feedback. *Geophysical Research Letters* 34: L06604, doi:06610.01029/02006GL029021.

———. 2008. A century of warming in Lake Superior. *Limnology and Oceanography* 53:2724–2730.

Balanyá, J., J.M. Oller, R.B. Huey, G.W. Gilchrist, and L. Serra. 2006. Global genetic change tracks global climate warming in *Drosophila subobscura. Science* 313:1773–1775.

Berthold, P. 1996. *Control of Bird Migration*. London: Chapman and Hall.

Both, C., and Visser, M.E. 2001. Adjustment to climate change is constrained by arrival date in a long-distance migrant bird. *Nature* 411:296–298.

Both, C., M. van Asch, R.G. Bijlsma, A.B. van den Burg, and M.E. Visser. 2009. Climate change and unequal phenological changes across four trophic levels: Constraints or adaptations? *Journal of Animal Ecology* 78:73–83.

Bradley, N.L., A.C. Leopold, J. Ross, and H. Wellington. 1999. Phenological changes reflect climate change in Wisconsin. *Proceedings of the National Academy of Sciences* 96:9701–9704.

Bradshaw, W.E., S. Fujiyama, and C.M. Holzapfel. 2000. Adaptation to the thermal climate of North America by the pitcher-plant mosquito, *Wyeomyia smithii. Ecology* 81:1262–1272.

Bradshaw, W.E., and C.M. Holzapfel. 2008. Genetic response to rapid climate change: It's seasonal timing that matters. *Molecular Ecology* 17:157–166.

Brocke, R. 1970. The winter ecology and bioenergetics of the opossum, *Didelphis marsupialis*, as distributional factors in Michigan. PhD dissertation, Michigan State University, East Lansing, MI.

Brown, C.R. and M.B. Brown. 2000. Weather-mediated natural selection on arrival time in cliff swallows (*Petrochelidon pyrrhonata*). *Behavioral Ecology and Sociobiology* 47:339–345.

Brown, J.H. 1995. *Macroecology.* Chicago: University of Chicago Press.

CCSP. 2009. *Global Climate Change Impacts in the United States.* U.S. Climate Change Science Program, Unified Synthesis Product. U.S. Climate Science Program, Washington, DC. http://www.globalchange.gov/publications/reports/scientific-assessments/us-impacts.

Christensen, J.H., B. Hewitson, A. Busuioc et al. 2007. Regional climate projections.

In *Climate Change 2007: The Physical Science Basis*, ed. by S. Solomon, D. Qin, M. Manning et al., 847–940. Cambridge: Cambridge University Press.

ClimateWizard. 2009. ClimateWizard is a web-based tool for making past and projected climate data from various sources more accessible, and was created through a collaboration between The Nature Conservancy, the University of Washington, and the University of Southern Mississippi (http://www .climatewizard.org). The PRISM data shown here were created February 4, 2007, and were processed with ClimateWizard in August 2009.

Desai, A.R., J.A. Austin, V. Bennington, and G.A. McKinley. 2009. Stronger winds over a large lake in response to weakening air-to-lake temperature gradient. *Nature Geoscience*: doi: 10.1038/NGE0693.

Dobiesz, N.E., and N.P. Lester. 2009. Changes in mid-summer water temperature and clarity across the Great Lakes between 1968 and 2002. *Journal of Great Lakes Research* 35:371–384.

Dobson, A. 2009. Climate variability, global change, immunity, and the dynamics of infectious diseases. *Ecology* 90:920–927.

Donner, D.M., J.R. Probst, and C.A. Ribic. 2008. Influence of habitat amount, arrangement, and use on population trend estimates of male Kirtland's warblers. *Landscape Ecology* 23:467–480.

Dunn, P.O., and D.W. Winkler. 1999. Climate change has affected the breeding date of tree swallows throughout North America. *Proceedings of the Royal Society of London Series B–Biological Sciences* 266:2487–2490.

Etges, W.J., and M. Levitan. 2008. Variable evolutionary response to regional climate change in a polymorphic species. *Biological Journal of the Linnean Society* 95:702–718.

Etterson, J.R., and R.G. Shaw. 2001. Constraint to adaptive evolution in response to global warming. *Science* 294:151–154.

Ewert, D.N., and M.J. Hamas. 1995. Ecology of migratory landbirds during migration in the Midwest. In *Management of Midwestern Landscapes for the Conservation of Neotropical Migratory Birds*, ed. by F.R. Thompson III, 200–208. U.S. Dept. Agriculture, Forest Service, North Central Forest Experiment Station, Gen. Tech. Rep. NC-187.

Field, C.B., L.D. Mortsch, M. Brklacich et al. 2007. North America. In *Climate Change 2007: Impacts, Adaptation and Vulnerability*, ed. by M.L. Parry, O.F. Canziani, J.P. Palutikof, P.J. van der Linden, and C.E. Hanson, 617–652. Cambridge: Cambridge University Press.

Foden, W., G. Mace, J.-C. Vié et al. 2008. Species susceptibility to climate change impacts. In *The 2008 Review of the IUCN Red List of Threatened Species*, ed. by J.-C. Vié, C. Hilton-Taylor, and S.N. Stuart. Gland, Switzerland: IUCN.

Gibbs, J.P., and A.R. Breisch. 2001. Climate warming and calling phenology of frogs near Ithaca, New York, 1900–1999. *Conservation Biology* 15:1175–1178.

Gibson, W.P., C. Daly, T. Kittel et al. 2002. Development of a 103-year high-resolution climate data set for the conterminous United States. In *Proceedings, American Meteorological Society, Portland, OR, May 13–16*, 181–183. http://www.prism.oregonstate.edu/pub/prism/docs/appclim02103yr_hires_dataset-gibson.pdf.

Gienapp, P., C. Teplitsky, J.S. Alho, J.A. Mills, and J. Merilä. 2008. Climate change and evolution: Disentangling environmental and genetic responses. *Molecular Ecology* 17:167–178.

Girvetz, E.H., C. Zganjar, G.T. Raber, E.P. Maurer, P. Kareiva, and J.J. Lawler. 2009. Applied climate-change analysis: The Climate Wizard tool. PLoS ONE 4:e8320. doi:8310.1371/journal.pone.0008320.

Groisman, P.Y., R.W. Knight, T.R. Karl, D.R. Easterling, B. Sun, and J.H. Lawrimore. 2004. Contemporary changes of the hydrological cycle over the contiguous United States, trends derived from *in situ* observations. *Journal of Hydrometeorology* 5:64–85.

Gu, L., P.J. Hanson, W. Mac Post et al. 2008. The 2007 eastern US spring freezes: Increased cold damage in a warming world? *Bioscience* 58:253–262.

Hayhoe, K., J. VanDorn, T. Croley II, N. Schlegal, and D. Wuebbles. 2010a. Regional climate change projections for Chicago and the US Great Lakes. Journal of Great Lakes Research 36 (Suppl. 2):7–21.

Hayhoe, K., S. Sheridan, L. Kalkstein, and S. Greene. 2010b. Climate change, heat waves, and mortality projections for Chicago. *Journal of Great Lakes Research* 36 (Suppl. 2):65–73.

Holt, R.D. 1990. The microevolutionary consequences of climate change. *Trends in Ecology and Evolution* 5:311–315.

Hong, Y., A. Steinman, B. Biddanda, R. Rediske, and G. Fahnenstiel. 2006. Occurrence of the toxin-producing cyanobacterium *Cylindrospermopsis raciborskii* in Mona and Muskegon Lakes, Michigan. *Journal of Great Lakes Research* 32:645–652.

Jones, M.L., B.J. Shuter, Y. Zhao, and J.D. Stockwell. 2006. Forecasting effects of climate change on Great Lakes fisheries: Models that link habitat supply to population dynamics can help. *Canadian Journal of Fisheries and Aquatic Sciences* 63:457–468.

Kearney, M., R. Shine, and W.P. Porter. 2009. The potential for behavioral thermoregulation to buffer "cold-blooded" animals against climate warming. *Proceedings of the National Academy of Sciences* 106:3835–3840.

Kutz, S.J., E.P. Hoberg, L. Polley, and E.J. Jenkins. 2005. Global warming is changing the dynamics of Arctic host-parasite systems. *Proceedings of the Royal Society B* 272:2571–2576.

Levitan, M. 2003. Climatic factors and increased frequencies of "southern" chromosome forms in natural populations of *Drosophila robusta*. *Evolutionary Ecology Research* 5:597–604.

Magnuson, J.J., K.E. Webster, R.A. Assel et al. 1997. Potential effects of climate changes on aquatic systems: Laurentian Great Lakes and Precambrian Shield Region. *Hydrological Processes* 11:825–871.

Marra, P.P., C.M. Francis, R.S. Mulvihill, and F.R. Moore. 2005. The influence of climate on the timing and rate of spring bird migration. *Oecologia* 142:307–315.

McCormick, M.J., and G.L. Fahnenstiel. 1999. Recent climatic trends in nearshore water temperatures in the St. Lawrence Great Lakes. *Limnology and Oceanography* 44:530–540.

Meehl, G.A., T.F. Stocker, W.D. Collins et al. 2007. Global climate projections. In *Climate Change 2007: The Physical Science Basis*, ed. by S. Solomon, D. Qin, M. Manning et al., 747–843. Cambridge and New York: Cambridge University Press.

Millien, V., S.K. Lyons, L. Olson, F. Smith, A.B. Wilson, and Y. Yom-Tov. 2006. Ecotypic variation in the context of global climate change: Revisiting the rules. *Ecology Letters* 9:853–869.

Moore, F.R., R.J. Smith, and R. Sandberg. 2005. Stopover ecology of intercontinental migrants: En route problems and consequences for reproductive performance. In *Birds of Two Worlds*, ed. by R. Greenberg and P.P. Marra, 251–261. Washington, DC: Smithsonian Institution Press.

Murray, D.L., E.W. Cox, W.B. Ballard et al. 2006. Pathogens, nutrient deficiency, and climate influences on a declining moose population. *Wildlife Monographs* 166:1–30.

Myers, P., B.L. Lundrigan, S.M.G. Hoffman, A.P. Haraminac, and S.H. Seto. 2009. Climate-induced changes in the small mammal communities of the Northern Great Lakes Region. *Global Change Biology* 15:1434–1454.

Ostfeld, R.S. 1997. The ecology of Lyme-disease risk. *American Scientist* 85:338–46.

Parmesan, C. 2006. Ecological and evolutionary responses to recent climate change. *Annual Review of Ecology, Evolution, and Systematics* 37:637–669.

Parmesan, C., and G. Yohe. 2003. A globally coherent fingerprint of climate change impacts across natural systems. *Nature* 421:37–42.

Price, J., S. Droege, and A. Price. 1995. *The Summer Atlas of North American Birds*. London/San Diego: Academic Press.

Reále, D., A.G. McAdam, S. Boutin, and D. Berteaux. 2003. Genetic and plastic responses of a northern mammal to climate change. *Proceedings of the Royal Society of London Series B–Biological Sciences* 270:591–596.

Root, T.L. 1988. Environmental factors associated with avian distributional boundaries. *Journal of Biogeography* 15:489–505.

Root, T.L., J.T. Price, K.R. Hall, S.H. Schneider, C. Rosenzweig, and J.A. Pounds. 2003. Fingerprints of global warming on wild animals and plants. *Nature* 421:57–60.

Root, T.L., D.P. MacMynowski, M.D. Mastrandrea, and S.H. Schneider. 2005. Human modified temperatures induce species changes: Joint attribution. *Proceedings of the National Academy of Sciences* 102:7465–7469.

Rosenzweig, C., G. Casassa, D.J. Karoly et al. 2007. Assessment of observed changes and responses in natural and managed systems. In *Climate Change 2007: Impacts, Adaptation and Vulnerability*, ed. by M.L. Parry, O.F. Canziani, J.P. Palutikof, P.J. van der Linden, and C.E. Hanson, 79–131. Cambridge: Cambridge University Press.

Smith, R., M. Hamas, M. Dallman, and D. Ewert. 1998. Spatial variation in foraging of the Black-throated Green Warbler along the shoreline of northern Lake Huron. *Condor* 100:474–484.

Strode, P.K. 2003. Implications of climate change for North American wood warblers (*Parulidae*). *Global Change Biology* 9:1137–1144.

Thomas, C.D., E.J. Bodsworth, R.J. Wilson et al. 2001. Ecological and evolutionary processes at expanding range margins. *Nature* 411:577–581.

Thomas, C.D., A. Cameron, R.E. Green et al. 2004. Extinction risk from climate change. *Nature* 427:145–148.

Trenberth, K.E., P.D. Jones, P. Ambenje et al. 2007. Observations: Surface and atmospheric climate change. In *Climate Change 2007: The Physical Science Basis*, ed. by S. Solomon, D. Qin, M. Manning et al., 235–336. Cambridge and New York: Cambridge University Press.

Trumpickas, J., B.J. Shuter, and C.K. Minns. 2009. Forecasting impacts of climate change on Great Lakes surface water temperatures. *Journal of Great Lakes Research* 35:454–463.

Tucker, J.K., C.R. Dolan, J.T. Lamer, and E.A. Dustman. 2008. Climatic warming, sex ratios, and red-eared sliders (*Trachemys scripta elegans*) in Illinois. *Chelonian Conservation and Biology* 7:60–69.

Visser, M.E. 2008. Keeping up with a warming world: Assessing the rate of adaptation to climate change. *Proceedings of the Royal Society B–Biological Sciences* 275:649–659.

Visser, M.E., and C. Both. 2005. Shifts in phenology due to global climate change: The need for a yardstick. *Proceedings of the Royal Society B–Biological Sciences* 272:2561–2569.

Visser, M.E., L.J.M. Holleman, and P. Gienapp. 2006. Shifts in caterpillar biomass phenology due to climate change and its impact on the breeding biology of an insectivorous bird. *Oecologia* 147:164–172.

Vucetich, J.A., and R.O. Peterson. 2004. The influence of top-down, bottom-up, and abiotic factors on the moose (*Alces alces*) population of Isle Royale. *Proceedings of the Royal Society of London, Series B* 271:183–89.

Willis, C.G., B. Ruhfel, R.B. Primack, A.J. Miller-Rushing, and C.C. Davis. 2008. Phylogenetic patterns of species loss in Thoreau's woods are driven by climate change. *Proceedings of the National Academy of Sciences* 105:17029–17033.

Winder, M., and D.E. Schindler. 2004. Climate change uncouples trophic interactions in an aquatic ecosystem. *Ecology* 85:2100–2106

Wilmers, C.C., E. Post, R.O. Peterson, and J.A. Vucetich. 2006. Predator disease out-break modulates top-down, bottom-up and climatic effects on herbivore population dynamics. *Ecology Letters* 9:383–389.

PART TWO

Decision Making and Climate Change

Decision Making under Climate Uncertainty

The Power of Understanding Judgment and Decision Processes

SABINE M. MARX AND ELKE U. WEBER

THE DISCIPLINES OF ECONOMICS AND POLITICAL SCIENCE, AS WELL as applied climate science, have added a great deal to our understanding of the obstacles to the use of climate information. However, in order for climate information to be fully embraced and successfully implemented into risk management, the issue needs to be looked at in terms of risk communication to human decision makers—as individuals (e.g., a farmer in Ontario) and in groups (e.g., Chicago city council, tourism boards). What is special about human risk perception and decision making under risk and situations of uncertainty regarding climate? This is where psychology, as applied in behavioral economics and behavioral game theory, offers important insights and tools to design effective risk-management processes, which can be of potential use to adaptive decision making in the Great Lakes region. In this chapter, we first discuss uncertainty as a barrier to predictability. We review how normative and descriptive models differ in their postulates about the processes by which people predict the likelihood of uncertain events, and choose among actions with uncertain outcomes, and among actions with delayed outcomes. Throughout this paper, we discuss the challenges (and possibly opportunities) that arise from the fact that decision makers employ simplifying heuristics that take advantage of memory and past experience, but can also lead to systematic biases, and have multiple and oftentimes conflicting goals as they are influenced by a range of qualitatively different incentives in their judgments, decisions, and actions. The chapter concludes with suggestions on how to overcome barriers of uncertainty by using insights from behavioral decision research in constructive ways to design climate risk communication and effective decision environments that will be effective in achieving goals of possible policy interventions.

UNCERTAINTY AS A BARRIER TO PREDICTABILITY

Humans have a great need for predictability. It makes up an important part of our need for safety and security (Maslow 1943). Predictability has survival value. It provides control, helps avoid threats to physical and material well-being, and frees us from fear and anxiety. Furthermore, it allows us to plan and budget for the future. However, our abilities to predict the outcome of an action or event can be impaired when we are faced with uncertainty, i.e., situations where it is impossible to exactly describe the future outcomes of actions that are taken now. Uncertainty means that there may be unknown outcomes, unknown probabilities, and immeasurable components, leading to a real or perceived lack of control. While the term *risk* is used to describe choices where all outcomes and their probabilities are explicitly described, as in the case of choices between monetary lotteries in laboratory experiments ("decisions from description," discussed below), most real-world decisions do not provide this level of information, and involve uncertainty about possible consequences and their likelihoods.

Information about climate change and its impacts on the ecosystem and society, as well as mitigation and adaptation strategies, have a wide range of uncertainty associated with them. They range from model uncertainty (fundamental, structural, and parametric; see Easterling et al. in this volume) and technological uncertainty (will it work?) to social uncertainty (what will others do?), to name but a few.

Whether perceived or real, lack of control when confronted with climate change raises anxiety, individually and socially. Moderate levels of anxiety are desirable, because they motivate behaviors to regain control, observable in protective or evasive action to mitigate risk, information search, and theory building (see figure 1). Science and technology development themselves can be put into this category. The developments of forecasts for weather, climate, earthquakes, or financial markets are some of society's ways of reducing moderate levels of anxiety about sources of unpredictability.

The strength of our desire for control is evident in situations where our need for control is so strong that it leads to wishful thinking. We perceive an "illusion of control" (Langer 1975) in situations that are obviously

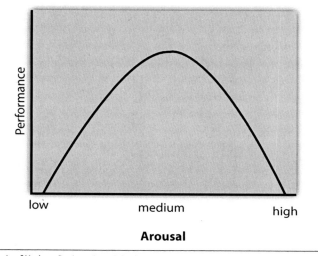

Figure 1. Graph of Yerkes–Dodson Law Principle

determined by chance, like predicting the color that the next spin of the roulette wheel will land on, based on past outcomes. This is often described as "gambler's fallacy" or probability matching (expecting heads from a coin flip after a run of tails). An increasing number of studies suggest that people tend to turn to superstitious or magic beliefs and strategies when in situations of uncertainty and stress (Case et al. 2004; Keinan 1994, 2002; Felson and Gmelch 1979). The illusion of control and superstition, although irrational from a scientific point of view, may be an adaptive response to an uncertain world (Haselton and Nettle 2006). Superstitious beliefs invoked during instances of uncontrollability may prevent or interrupt subsequent performance impairment (Dudley 1999). Recent studies in health care have shown beneficial physiological effects, such as pain relief (Wager et al. 2004; Thompson 1981; Wager, Scott, and Zubieta 2007), reduction of stress through illusory control (Alloy and Abramson 1982), and wound healing (Kiecolt-Glaser et al. 1995).

TWO SOURCES OF INFORMATION

To enable informed decision making under uncertainty, people have two forms of information available to them: 1) *description* of possible outcomes and their likelihoods provided by others, often provided in the form of a statistical summary, and 2) personal *experience*, typically acquired over time (for more discussion of learning, see Weber 2006; 2010). The former is illustrated by seasonal climate forecasts for the next growing season, or hurricane warnings issued by the National Hurricane Center based on scientific models. This information describes outcome distributions in the form of possible outcomes and their probabilities, provided numerically or graphically. Yet, we also often draw on knowledge acquired by personal exposure. While in decisions from experience, the outcome distribution may initially be unknown, repeated exposure and repeated choices provide us with knowledge of possible outcomes and their likelihood. It can, for example, lead farmers to make intuitive forecasts of climate in the next growing season based on years of past experience. Cash-crop farmers in Ontario take note that excessive rains in the spring tend to happen every five years, and base their farming decisions on the continuation of this pattern, without taking potential alterations due to climate change into account (Bradshaw et al., this volume). Similarly, repeated experience with warning signs and subsequent events leads people living in hurricane-prone areas to an intuitive assessment of the likelihood of being affected by a hurricane.

The processing of these two different forms of information (description and experience) involves partially overlapping but also distinct processing systems in the brain: the analytic system and the experiential/affective system, respectively (Damasio 1995; Chaiken and Trope 1999; Sloman 1996; Epstein 1994; Marx et al. 2007). The processing of descriptive statistical summary information and description-based decision making requires extensive use of the analytic processing system. Analytic processing is effortful and slow, and requires conscious awareness and knowledge of rules, e.g., probability calculus, Bayesian updating, formal logic. It is activated during conscious, calculation-based decisions.

Experience-based decision making relies more on regions of the brain involved in associative learning, the experiential/affective processing system. This system operates fast and automatically. Experiential processing relates current situations to memories of one's own experiences, or those of others that we may have learned about. Experiential knowledge associates behavior and consequences and is acquired through trial-and-error learning. There is an emphasis on decision outcomes and probabilities not explicitly represented. Because past experiences often evoke strong feelings, emotions are a powerful class of associations. Risks are represented as a "feeling" that serves as an "early warning system" (Loewenstein et al. 2001). Strong emotions, such as pleasure and pain, fear, anger, horror, joy, and awe, associated with past events make the experience more memorable and therefore often dominant in processing (Slovic et al. 2002; Slovic et al. 2007).

The two processing systems operate in parallel and interact to some extent. However, if the output of the two systems is in conflict, behavior is typically determined by the experiential/affective processing system, because it is faster, delivers output earlier, and is very vivid.

The discrepancy in output of the two systems often accounts for public controversies and debates about the magnitude and acceptability of risks, as in the case of nuclear power or genetic engineering. Technical experts and academics tend to rely more heavily on analytic processing in their definition and evaluation of possible risks, while politicians, end-user stakeholders, and the general public rely more heavily on experiential/affective processing. This is one reason why often, climate science information does not readily feed into decision makers' existing decision models and procedures, which Moser (this volume) describes as problems of compatibility of climate science output and decision makers' decision models and processes. Translation of science into usable knowledge can be improved by a two-way dialogue and information sharing between scientists and decision makers about uncertainties (scientists' knowledge of statistical probabilities and farmers' experiences with probabilistic events), as suggested by Easterling et al. (this volume).

NORMATIVE MODELS FOR DECISION MAKING UNDER UNCERTAINTY

Scholars have taken two general paths in their work on decision making under conditions of uncertainty: *normative* and *descriptive* models. While descriptive models attempt to describe how people make decisions in the real world, normative models (e.g., Expected Utility Theory (EUT) and Bayesian Updating) prescribe how judgments or decisions ought to be made.

Normative approaches to decision making address what will happen in the future and how likely it is. Normative decision theory assumes that each decision is decomposed into four components: (1) a set of possible actions, (2) a set of possible future states of the world, (3) information on the probability of different future states of the world, and (4) information about the outcomes of possible actions under future states of the world. Rational predictions based on models of contributing processes offer some good answers to the following questions:

What will happen?

Classical normative approaches assume that there is an observation history of past outcomes y going back in time t (y_t, y_{t-1}, y_{t-2}, ... y_{t-n}) that are used to predict future outcomes using regression techniques (y_{t+1}, y_{t+2}, ... y_{t+n}), including time series analysis.

How likely is it?

Classical normative decision models assume separable outcome and probability information. Probabilities can be interpreted on the basis of relative frequencies of events in repeated trials, or as degrees of belief or confidence of a proposition. Based on a normative algorithm called the Bayesian theorem, new pieces of evidence can be incorporated into prior beliefs.

When will it happen?

Traditionally, normative choice models do not include a temporal component, yet in the context of climate variability and climate change, time matters. Climate change predictions usually have long multi-period time horizons. A prominent approach for making long-range predictions, for instance until 2100 and beyond, is the *scenario technique*. Combining different parameter constellations (highly, medium, less probable) and differing assumptions about demographic change, consumption patterns, etc., this technique is applied by the Intergovernmental Panel on Climate Change (IPCC), e.g., the IPCC Special Report on Emission Scenarios, IPCC Assessment Reports, and US Science Assessment Products (IPCC 2001a, 2001b, 2007; SAP/USP 2009). Given a set of assumptions about future developments of explanatory variables, and given a model for the relationship between explanatory and explained variables, the time path of the explained variable is predicted (embedded in confidence bands). Different time paths can be established for the different sets of assumptions.

Where will it happen?

Traditionally, normative models do not include a spatial component, yet in the context of climate variability and change, geography or space is of great importance. For instance, we want to know where precipitation patterns will change most in a region (Andreson, this volume) or where decreases in lake-water levels will have the greatest impact in the Great Lakes (Mackey, this volume). Geographic Information System (GIS) models of past and present situation-specific areas of the world are combined with spatial models to make spatial predictions (again embedded in confidence bands). The importance and value of regional scenarios for the Great Lakes is demonstrated by the Pileus Project and the Lake St. Clair Project, described by Winkler et al. and by Mackey, respectively, in this volume.

By taking available alternatives to what the future states of the world and their probabilities could be, and what outcomes the different choice alternatives would have under different future states of the world, normative models offer many advantages: (1) they tell us how a rational decision maker should behave, which provides a good benchmark against which actual behaviors can be compared; (2) they have a clear analytic basis that can easily be updated; and (3) the cost and benefit of generation/acquisition of additional information can easily be assessed.

Normative models, however, also have several weaknesses: (1) they do not explain why or how people make the decisions we observe; (2) they expect decision makers to be fully rational; (3) they assume the decision maker to be well informed about key components of the decision problem; (4) they often assume that there is sufficient knowledge about the outcomes; (5) they assume knowledge of quantitative probabilities and ignore the use of affective reactions and other heuristic processes to assess likelihood; (6) they often consider only one individual decision maker and not groups; (7) they do not consider the interactive process and the interactions evolving while decisions are made; (8) utility functions of decision makers are difficult to identify; and (9) they often lack temporal and spatial components, which matter greatly to impacts of climate variability and change.

In sum, normative prediction approaches are important. They help structure information requirements, help structure the decision situation, and can serve as relatively simple benchmarks. Yet, in order to make reliable predictions of how people actually decide, descriptive models, which integrate psychological processes into the decision model, are needed as supplements.

DESCRIPTIVE MODELS OF DECISION MAKING UNDER UNCERTAINTY

Normative models have great advantages when dealing with probability updating and deep uncertainty, as discussed in this volume's chapter by Easterling et al. Real-life observations, however, indicate that people do not behave fully rationally, but instead take shortcuts in their processing by applying heuristics. Theoretical approaches that try to model how people actually make decisions are called descriptive, and include Prospect Theory, Theory of Constructed Choice, Theory of Context-Dependent Choice, and others. Descriptions of decision making take into consideration the multiple modes by which people have been observed to make decisions, namely calculation-based, rule-based, and affect-based decisions (Weber and Lindemann 2007). They assume a combination of analytic and experiential processing of information, and take into account the different effects that personal experience and statistical summary information have on the assessment and management of risks (Marx et al. 2007). In the remainder of this section, we address two phenomena that play out strongly in the context of climate change communication, and decision making under climate uncertainty: (1) people are overconfident in the accuracy of a prediction or decision, and (2) deterministic, causal, and experiential/affective thinking is more prevalent than statistical and probabilistic thinking. Experiential processing employed to predict uncertain events makes use of the following heuristics, which utilize stored past experience.

Overconfidence

The *overconfidence bias* was first described by Alpert and Raiffa (Alpert and Raiffa 1969) and is defined as the tendency of individuals to overestimate the preciseness of their knowledge, and to express excessive optimism concerning the probability of a certain favorable/unfavorable outcome in the future. While overconfidence is often studied using lay people, it is just as commonly found among scientists and other experts. For example, Robert Milikan, who was awarded a Nobel Prize in Physics in 1923, once said, "There is no likelihood man can ever tap the power of the atom," and Lord Kelvin, who served as president of the Royal Science Society for five years, asserted that "Heavier than air flying machines are impossible."

People tend to be particularly overconfident in their judgments when it is difficult to make correct judgments. In a classic study by Lichtenstein and

Fischhoff (1977), people were presented with 12 children's drawings and asked whether they came from Europe or Asia. In a second task, participants were asked to estimate the probability that each of their judgments was right. Although only 53 percent of the judgments were accurate (essentially at chance level), people rated their confidence, on average, as 68 percent. The researchers conclude that in two-alternative judgments, the correspondence between accuracy and confidence is as follows: (1) overconfidence is greatest when accuracy is near chance levels; (2) overconfidence diminishes as accuracy increases from 50 to 80 percent, and once accuracy exceeds 80 percent, people often become underconfident; (3) the gap between accuracy and confidence is smallest when accuracy is around 80 percent, and it grows larger as accuracy departs from this level; and (4) degree of overconfidence is greater the more difficult the task (Lichtenstein and Fischhoff 1977). Over the last three decades, many studies have replicated these findings for more commonplace judgments and predictions about behavior in contexts that were much more tailored to the expertise of the respective participant populations (Dunning et al. 1990; Vallone, Griffin, and Lin 1990).

It needs to be pointed out that low degrees of calibration (discrepancies between accuracy and confidence) are not linked to a decision maker's intelligence. Russo and Schoemaker (Russo and Schoemaker 1992) conducted experiments with high-level managers in many industries by asking ten questions tailored to the industry. Respondents were asked to provide a low and high estimate for each question, such that they were 90 percent certain the correct answer would fall within these limits, aiming for 90 percent hits and 10 percent errors. Typical outcomes fell in the range of 50 to 60 percent. That is not to say that people are *always* overconfident. There is some evidence that expert bridge players, bookies, and weather forecasters exhibit little to no overconfidence, most likely because they receive almost instantaneous and frequent feedback following their judgments (Lichtenstein, Fischhoff, and Phillips 1982; Murphy and Brown 1984; Murphy and Winkler 1984; Keren 1987).

There are multiple reasons for overconfidence. Overconfidence can be the result of selective information and memory search. Humans tend to seek or interpret new information in ways that confirm one's beliefs and avoid contradictory interpretations (*confirmation bias*) (Koriat, Lichtenstein, and Fischhoff 1980). There also are motivational reasons. We see a need to appear competent and confident to others and to ourselves.

In conclusion, the downside of confidence exceeding prediction accuracy is that it prevents people from seeking additional information and cues, thus preventing them from considering alternative judgments or decisions. There are also important implications for veridicality of personal recollections of climate information. On the positive side, confidence and optimism help to get the job done. The optimal balance is that one is somewhat "schizophrenic"—accurately assessing the correct probabilities of the success of a plan or decision under consideration and developing contingency plans ("Plan Bs") in the case of lower success probabilities. Once this is done, one can and should plunge ahead with Plan A with full optimism and enthusiasm.

Availability Heuristic

The availability heuristic allows people to make likelihood predictions based on what they remember, how easily these memories are retrieved, and how readily available those memories are. Ease of recall serves as indicator of likelihood. People have been found to employ the availability heuristic when asked for probability or frequency judgments, often of a comparative type (Tversky and Kahneman 1973). When people are asked to judge whether the probability of a blizzard is greater for November or January, they will try to recall storms that they remember occurring in either November or January ("on Thanksgiving weekend," or "around my birthday right after New Year's Day"). Whichever category provides more available concrete examples, or for which it feels easier to generate examples, is the one that is judged to be more likely. This rule of thumb works fairly well, because common events are easier to remember than uncommon ones. Yet not all easily recalled events are very likely. Some events are more available not because they occur more frequently, but because they have taken place more recently (*recency effect*); because they have been distorted by the media, which tends to over-report catastrophic rather than chronic risks; or because they are associated with strong emotions (*affect heuristic*). The availability heuristic can play a large role in judging the probabilities of extreme climate events (for instance drought, or an abnormally wet spring), because people can commonly recall unusually good or bad seasons. The response to long-term climate change information is different because most of us, unless we live in Alaska or other places already similarly affected by climate change, have not (yet) experienced the impacts associated with climate change, and cannot bring examples (whether frightening or pleasant) to mind. The availability heuristic makes us thus assume that the future will be similar to what we have experienced so far (Sunstein 2006).

Recency Effect

Experiential processing gives a lot of weight to recent observations. Since rare events have generally not occurred recently, they are underweighted (Weber, Shafir, and Blais 2004). Such underestimation of the risks of rare events based on experiential processing may contribute, for example, to the neglect of flood-control infrastructure by the federal government in recent decades. However, recency weighting also predicts that, if the statistically rare event *has* occurred in the very recent past, people will overreact to it (Hertwig et al. 2004). This makes decisions from personal experience far more volatile than decisions based on analytic processing of statistical information. The devastation caused by the Asian tsunami or the destruction due to Hurricane Katrina in the Gulf of Mexico are examples where recent rare events may have led people to overestimate the likelihood of subsequent similar events. In summary, when people base their decisions on personal experience with a risky option, recent outcomes strongly influence the evaluation of this risky option. As a result, low-probability events generate more concern than they deserve in those instances where they do occur, and less concern than their probability warrants when they haven't occurred in the recent past. Parallel to individual reactions, the media also have a tendency to over-report catastrophic rather than chronic risks, once they have occurred.

Affect Heuristic

Slovic et al. (2007) suggest that the biases in probability and frequency judgment that have been attributed to the availability heuristic may be due, at least in part, to affect. People turn to their emotions; they consult or refer to a repertoire of positive and negative images associated with the object or decision at hand. Affect is linked to risk perception in many ways. Slovic and colleagues have shown that feelings of dread are linked with perceptions of risk by way of risks being seen as uncontrollable, involuntary, inequitable, catastrophic, fatal, etc. (Slovic, Fischhoff, and Lichtenstein 1980).

Psychological Dimensions of Risk

Psychological risk dimensions (Fischhoff et al. 1978; Slovic, Fischhoff, and Lichtenstein 1986) address the affective reactions to risky choice options that are the determinants of perceived risk. This psychometric paradigm was established through psychological multidimensional scaling and multivariate analysis techniques to identify the characteristics of hazards that affect people's subjective feelings of being at risk. Affective risk dimensions influence judgments of the riskiness of material, physical, and environmental risks in ways that go beyond their objective consequences.

Based on results from multiple international studies in which people judged diverse sets of hazards (Slovic 1997), the various hazards were reduced and combined into two factors: (1) dread risk and (2) unknown risk. Items grouped under the "dread risk" factor trigger our early (physical) warning systems, such as speeding up our heart rate and producing anxiety (e.g., nuclear war or nuclear-reactor accidents). This factor addresses whether a hazard is controllable/uncontrollable, dreaded/not dreaded, equitable/not equitable, of individual/global catastrophic impact, easily/not easily reduced, risk decreasing/increasing, voluntary/involuntary, bears fatal/non-fatal consequences, or poses low/high risk to future generations.

The second factor refers to the degree to which a hazard is observable/unobservable, an old/new risk, known/unknown to those exposed, known/unknown to science, and whether it has a delayed/immediate effect. Examples of items perceived as very high risk are newer technologies such as GMO technology, which may have unforeseen consequences.

Weber has looked more closely at climate change within this two-factor space (Weber 2006). Following Slovic's psychometric paradigm, the psychological risk dimensions associated with climate change predict the level of alarm or worry. Assuming that people perceive climate change as a gradual process of temperature and precipitation change, and a gradual change in the frequency and severity of extreme events such as hurricanes, cold spells, or heat waves, the risk would appear to be well-known and controllable (if illusory) to those who can afford to move to safer ground. Because the effect of climate change is perceived as delayed, many people do not (yet) perceive it as a threat. When people base their decisions on statistical descriptions about the hazards of climate change, the time-delayed and abstract nature of the risk does not evoke strong visceral reactions. If climate change is depicted as abrupt/rapid and catastrophic, as in the movie "The Day After Tomorrow," the potential for raising a visceral reaction to the risk is much greater (Leiserowitz 2004, 2006). As a result of cognitive biases, distorted media coverage, and misguiding experience, people tend to neglect uncertainty, and under- or overestimate risks, depending on circumstances.

DESCRIPTIVE CRITIQUES OF NORMATIVE MODELS

This section first discusses two normative models commonly used to predict decision making under uncertainty, and then introduces psychological extensions used to describe how and why people behave differently than they ought to.

Expected Utility Theory (EUT)

Facing uncertainty, EUT asserts that decision makers should choose between two options by comparing their expected utility values (the weighted sums of the utility values of outcomes multiplied by their respective anticipated probabilities): $U = \Sigma_i\ p_i u(x_i)$, where "i" means states of the world; in each state of the world (i) the individual receives x_i dollars; the probability of receiving x_i is p_i. People should behave as if they were maximizing the expected utility of choice options. The shape of the utility function serves as a measure of risk aversion (concave) and risk seeking (convex). Risk-neutral individuals have linear utility functions.

Many individuals do not seem to have consistent utility functions in the face of risk. A variation of EUT, Subjective Expected Utility Theory (SEUT) allows for a personal utility function and subjective, or personal, probabilities. SEUT can also be applied to problems where objective probabilities cannot be determined in advance, for instance in cases where the outcome will only occur once.

A psychologically more realistic alternative to expected utility is provided by Prospect Theory. Tversky and Kahnemann (1979, 1992) modified EUT with a utility, or value function, that is defined over gains and losses compared to a reference point (instead of over absolute wealth). The value function is an S-shaped curve with a point of inflection at the reference point. Manipulations of what people use as their reference point determine how outcomes get framed (e.g., as relative gains or relative losses). Reference-point framing matters because the prospect theory value function is concave for gains (risk averse), convex for losses (risk seeking). It is also steeper for losses than for gains, i.e., losses loom larger than gains of the same amount (loss aversion). Prospect theory predicts the following risk attitudes as the result of diminishing marginal sensitivity to both gains and losses: risk aversion over prospects involving gains, and risk seeking for prospects involving losses. However, due to nonlinear probability weighing, where people tend to overweigh small probabilities, it also predicts risk aversion for small-probability losses, which explains why people buy insurance, and risk seeking for small probability gains, which explains why people buy lottery tickets.

Discounted Utility Theory

The standard normative approach to evaluating possible future consequences of an action is to discount them based on their time delay, meaning that consequences are considered to be less important with each year they are delayed. The utility of an outcome is discounted as a function of its time delay, assuming a constant discount rate for all time periods. Prevailing interest rates serve as a reasonable standard for discounting (the reduction of value per year of delay), and all future outcomes should be discounted at a continuously compounded, exponential rate.

Empirical findings show a wide range of deviations from discounted utility theory. In reality, people are impatient and tend to demand much greater premiums for delay when immediate consumption is an option than in situations where all possible outcomes lie in the future (Mischel, Grusec, and Masters 1969; Read 2001). Thaler found that people discount gains more than losses and small outcomes more than large ones (Thaler 1981). Studies by Loewenstein and Weber et al. show greater discounting when delaying rather than accelerating consumption (Weber et al. 2007; Loewenstein 1988). Furthermore, multiple outcomes are discounted differently than single outcomes. Weber and Chapman have found evidence for a combined effect of delay and uncertainty, where decision makers equate uncertainty and delay (Weber and Chapman 2005). Several theoretical formulations have attempted to explain these anomalies, such as models of hyperbolic discounting, awareness of future self, utility from anticipation, instantaneous utility function, habit-formulation models, reference-point models, projection bias, mental-accounting models, choice bracketing, multiple-self models, temptation utility, etc. (Frederick, Loewenstein, and O'Donoghue 2002, 2003). Yet another view, Query Theory, put forward by Weber et al., points out that the order in which the individual retrieves different classes of reasons for the decision greatly impacts the discount rate applied (Weber et al. 2007). In delay decisions, people focus first on arguments for the default option of immediate consumption, whereas in acceleration decisions, they focus first on the default option of later consumption. Weber and Johnson (2006) show that while the non-default option is also considered, argument retrieval is less successful for second queries. Weber and colleagues show that discounting future outcomes in delay decisions can be reduced greatly when decision makers are prompted to first generate reasons for waiting to get more later (Weber et al. 2007).

Environmental decisions, where both financial and nonfinancial outcomes are realized at different points in time, bear additional complications:

1. Discounting may be domain-dependent (Gattig and Hendrickx 2007). Chapman's studies show greater time discounting for health than for money, with large variability and poor correlation across individuals between discount rates for health and money outcomes. This is attributable to whether or not people think of health as tradable with money (Chapman 1996, 2002, 2003). Chapman noted lower discount rates when the outcome affected a client population rather than oneself. This could be interpreted as less discounting of the future for a social rather than an individual health goal. Treadwell found discount rates of zero for future health (Treadwell 1997).

2. Time horizons of environmental outcomes are often much larger than those typically studied with monetary and even health outcomes, e.g., intergenerational goals, or environmental enhancements that are imagined to endure indefinitely.

3. While the money in financial scenarios is usually received in the form of a lump sum, the environmental outcomes are likely to be experienced as a stream of benefits (or losses) spread out over time.

4. In contrast to the financial scenarios typically studied, environmental outcomes generally have consequences beyond the individual decision maker.

5. Environmental decisions can be associated with either (a) purely social goals, e.g., a livable global climate for future generations; or (b) individual (as well as social) aspects that must be pursued in cooperation with others, e.g., reducing local air pollution, coping with water shortage. And finally,

6. Environmental outcomes are typically more uncertain than monetary outcomes.

Few studies have examined intertemporal preferences for non-monetary outcomes. The question of discount rates for non-monetary goals has been addressed most systematically in the literature on health outcomes (Bos, Postma, and Annemans 2005; Chapman 1996, 2002, 2003), and only a handful have explicitly studied environmental outcomes. Baron showed that people were insensitive to the amount of delay in a study of discounting of environmental goods, with significantly lower discounting for longer time intervals than for shorter intervals (Baron 2000). Guyse et al.

examined preferences for temporal sequences of environmental and monetary outcomes and found that students preferred improving sequences for environmental outcomes (as opposed to those that worsened over time), contrary to the findings for monetary outcome sequences (Guyse, Keller, and Eppel 2002). Studies by Hardisty and Weber (2009) indicate that over a short time period (1 year and 10 years), when personal consequences are salient, environmental outcomes are discounted similarly to monetary outcomes. Yet health outcomes were valued differently from money and the environment. In Hardisty's studies, discount rates varied more by valence (gains vs. losses) than by context. Context-specific discounting should be examined in a more systematic way using compatible scenarios for financial, environmental, health, and other outcomes. More research is also needed to determine the influence of longer time periods (20 years, 100 years) when social goals are more salient, and environmental/social outcomes may indeed show strong negative discount factors.

To address the special feature of (very) long-term environmental goals, a theory put forward by Krantz and Kunreuther (2007) departs from standard economic theories. According to their Context-Dependent Construction of Choice Theory (hereafter KK Theory), the benefit of an enduring enhancement to social or natural environment is not evaluated as a series of utility increments year by year—rather, it is perceived as a single goal pertaining to an indefinitely long time period. Since one is not integrating over time, there is no mathematical need for something such as exponential discounting. There can still be pure time preference: delaying an achievement of indefinite subsequent duration may change the value assigned to the goal. Yet such discounting may be slight, and there may even be negative discounting for some long-term goals and some values of delay.

Both normative and descriptive models ought to include a decision maker's multiple goals and incentives, and need to be able to deal with groups of decision makers and their potentially conflicting objectives, rather than focusing on individual decision makers. The latter point is particularly useful when addressing environmental decisions under uncertainty, which are most commonly made by groups composed of individuals who may have both aligned and conflicting goals. Not knowing what others' goals are and how they will behave further increases uncertainty.

Multiattribute Utility Theory

Multiattribute utility theory (MAUT) offers a structured approach designed to handle the tradeoffs among multiple objectives by assigning utility to outcomes on each dimension and weighting them by the relative importance of each dimension. MAUT is mostly used to evaluate choice among several decision alternatives in the presence of multiple goals (Debreu 1959; Krantz et al. 1971; Keeney and Raiffa 1976).

MAUT has commonly been applied in public-sector decisions and public-policy issues that require reconciliation of many interest groups, such as power plant–related decisions or military innovations, involving tradeoffs of cost, durability, lethality, and survivability. MAUT aims to separate utility into attributes. For instance, when choosing a medical treatment such as hormone replacement therapy (HRT), two attributes (with corresponding goals) to consider are long-term health and quality of life. Each of these attributes has several factors. Long-term health considerations include osteoporosis, heart disease, and breast-cancer concerns. The attributes that factor into quality of life are menopausal symptoms, side effects of the treatment, youthfulness, and so on. To use an environmental example, let us look at the adoption of a climate change policy that would reduce greenhouse gas emissions in a manner that would stabilize atmospheric concentrations at a specific value. The first attribute, short-term concerns, includes sacrifices such as money spent, perceived loss of standard of living and associated quality of life, and inequities. The second attribute, long-term benefits, consists of CO_2 emissions avoided, consequently averting dangerous climate impacts, saving lives and species, and helping the long-term well-being of the planet. To give an example from the agricultural sector, a farmer might be looking at the choice between different planting and cultivation regimens of a crop like soybeans that differ in terms of input costs (cost of seed, cost of necessary fertilizer, cultivation labor, etc.) and in terms of potential benefits (yields, time to harvest, etc.) under different climate conditions (Letson et al. 2009). For each level in this hierarchy, every factor is weighted by assigning points, which are then distributed among factors within a given group to indicate the relative importance of that factor in the decision to use or not use the treatment/to adopt or not adopt the policy. This method allows one to calculate a net weighted utility score for each factor, as well as a composite utility score, by adding the scores for the decision factors in the model. There are tradeoffs between the goals in our examples, long-term health and quality of life/short-term sacrifices and long-term benefits, plus the many other goals in the same decision. MAUT not only considers all of them, but considers them all in a consistent way, and all at once.

Multiattribute utility theory is based on the assumption that the fundamental normative principles of choice, such as transitivity, apply, and that decision makers are able to measure utility for combinations of different packages of goals within various outcomes. However, empirical evidence has shown that both of these assumptions are violated systematically and predictably.

Normative theory stipulates that our goals should be independent of context, yet context is often a main reason for violation of MAUT principles of transitivity and independence. A decision maker may prefer choice A over B in one context, B over C in another, and C over A in yet a third context leading to intransitivity. Context influences goals and also impacts how goals are bundled and thus how utility is assigned to various outcomes.

Building on Constructed Choice Theory (Lichtenstein and Slovic 2006; Slovic 1995), which states that people's preferences are not stable, but rather constructed with the decision context, Krantz and Kunreuther's Theory of Context-Dependent Constructed Choice assumes that decision makers choose among competing plans, each of which is designed to achieve multiple goals (Krantz and Kunreuther 2007). Context influences (1) which goals are active, i.e., different domains (financial, environmental, health, etc.) prime different goals (material, moral-ethical, social, etc.) with different intertemporal preferences; (2) which resources are available to achieve goals; and (3) which decision rules are considered. Besides context dependence, KK Theory rejects the notion that all goals merely contribute to a single utility. KK theory facilitates consideration of multiple types of goals, including emotional, social, environmental, and economic, as well as temporal-sequence goals (spreading positive experiences out over time vs. experiencing them all immediately [Loewenstein and Prelec 1993]).

This theory has many advantages over traditional multiattribute utility theory. While originally developed in the context of protective (insurance) decisions, Krantz and colleagues are currently working on an application to environmental decisions where context matters insofar as that environmental goals may be activated by new concepts (e.g., carbon sequestration) or by a cooperative plan to which others seem prepared to commit (e.g., the Kyoto Treaty).

Some decisions require tradeoffs that people think should not be made, such as trading off human life (or animal species, as in discussed in the chapter by Hall and Root [this volume]) for something else. In these cases, people often use non-compensatory (non-tradeoff) choice processes that avoid conflict (Hogarth 1987; Payne, Bettman, and Johnson 1993); or at least they avoid the realization of conflict by editing out elements that remind them of goal conflicts. Decision makers do not admit tradeoffs between the relevant attributes of the choice alternative; they eliminate alternatives without examining all attributes; they make decisions on an attribute-by-attribute basis; and separate utilities are not combined into a single utility value. Some of the more commonly studied non-compensatory, simplifying heuristics include elimination-by-aspects, acceptance-by-aspects, and lexicographic-by-features rules.

Two further complications arise in the context of conflicting goals. The difficulty of making tradeoffs often leads to decision inertia, and we end up with the status quo. Conflicting goals also lend themselves to strategic use of uncertainty, exemplified by the argument of wind power vs. species protection (birds flying into the wings of wind turbines).

Our goals include not only the selfish economic and material goals of *homo economicus*, but also social goals, emotional goals, and environmental goals. An additional complicating factor is that social goals are not easily quantifiable, making it even harder for decision makers to carry out the complex context-dependent measurements of utility required when multiple and often implicit goals are bundled together.

In sum, tradeoffs among choices are really tradeoffs between goals, which in turn are neither objective nor stable, but context dependent, subjective, and changing.

ECONOMIC AND OTHER INCENTIVES

Most decisions surrounding climate change mitigation and adaptation involve common pool resource dilemmas. While the normative perspective of economics, game theory, and behavioral research suggests bleak prospects for decisions that would sustain the earth's common resources, the picture might not be as bleak as it seems. The "tragedy of the commons" (Hardin 1968) might be better described as a drama, as a tragic outcome is by no means a foregone conclusion (Ostrom et al. 2002).

A better appreciation of the multiple ways in which people can look at information (framing, mental accounting), set goals (individual vs. social goals), and decide upon a course of action (using habits, rules, roles, affect, and calculations) can suggest ways in which environmentally impactful behavior can be presented to increase the likelihood for people to act in more collective ways, which in turn increases long-term individual benefits. The same insights can also guide the design and presentation of environmental policy options in those cases where intervention is considered necessary.

Cooperation can be facilitated by appealing to the social identity of people. Besides economic well-being, social affiliation and social approval are powerful human needs. Modification of economic incentives is difficult and expensive, yet the priming of social goals by the way situations are described or "framed" is often more (cost) effective and more feasible. For instance, giving a group a bonding task prior to the actual decision-making process can imply that the goal in question is a communal goal. Arora et al. (2012) show that cooperation in a social-dilemma game increased when the preceding task required cooperation. Even more subtle stimulations of group affiliation, such as a sticker attached to people's experiment materials to identify people as members of one group over another, can create group identity and subsequent cooperation (Brewer 2001; Brewer and Kramer 1986).

POLICY IMPLICATIONS

How can stakeholders' (public officials, members of the general public) attention be directed to climate change and variability? Most climate change information is presented in an analytic format; however, analytic appeals have not proven to be very effective in the past. Climate change predictions seem to contradict the experience of most people who lack personal exposure to actual climate change. Mitigative and adaptive actions often require immediate costs, sacrifices, and losses to achieve time-delayed benefits or gains, but both hyperbolic discounting and loss aversion argue against taking such actions.

Does this call for more emotional appeals? Is there wisdom in inducing people to worry more about climate change and variability? There are clearly good reasons to engage people's affective information-processing system by using visualization or graphic description of catastrophic climate change, or by making future changes concrete in simulations of conditions in local environments (Marx et al. 2007; Leiserowitz 2004, 2006). However, there are important caveats to be considered, namely the *finite pool of worry* and *single action bias*.

Finite Pool of Worry

The finite pool of worry refers to the fact that people have a limited budget to spend on worry (Linville and Fischer 1991). Several chapters in this volume address the multiple stressors under which decision makers in the Great Lakes operate. Easterling et al. and Bradshaw et al. (this volume) show that the success or failure of agricultural production/food security depends on a combination of climatic, disease-related, economic, political, and social variables. Scheraga (this volume) points at the myriad of pressures with which water-resource managers are confronted: besides the impacts of climate change, the redesigning of aging combined sewer systems for better control of sewer overflow need to take land-use change, land-cover change, and water pollution into account. For the city managers of Chicago, climate-change-related worries alone range from sewer-overflow protection and public-health infrastructure to protect the elderly against more frequent heat waves, to dredging of ports to counter the effects of water-level drops in Lake Michigan. As worry about one hazard increases, worry about other hazards decreases. For instance, research by Hansen, Marx, and Weber with farmers in Argentina showed that as worry about climate risks increased due to a forecast of La Niña conditions, farmers rated political situation and economic conditions as lower on the worry scale, even though political and economic risks had not changed since the climate forecast had been introduced (Hansen, Marx, and Weber 2004).

Single Action Bias

Single action bias refers to the tendency to engage in only a single corrective action to remove a perceived threat of a hazard, even when a whole range of responses is clearly advantageous. This occurs because the single action removes the "hazard flag." Radiologists have been found to detect a single abnormality, then stop searching for more, and therefore miss other potentially abnormal signs (Berbaum et al. 1998). Bradshaw et al. and Easterling et al. (this volume) describe in great detail the multiple actions and adaptive strategies (crop insurance, conservation tillage or zero tillage, choice of crop and seed variety, pest management, selection of planting dates, investment in irrigation, storage, diversification by adding livestock, acquisition of additional farmland, securing nonagricultural sources of income) necessary to deal with the multiple risks in agriculture (drought, excessive rain, cold and wet summers, heat, market stresses, political changes, etc.). Yet, a study of U.S. Midwestern agricultural decisions in the 1990s found that farmers engaged in only one activity to adapt to climate variability (*either* production practice, *or* pricing practice, *or* endorsement of government intervention) (Weber 1997). Similarly, farmers in the Argentine Pampas were less likely to use irrigation or crop insurance if they had capacity to store grain on their farms (Hansen, Marx, and Weber 2004). Decision-support tools such as checklists, or the EPA's Climate Assessment Tool for Water Resource Managers (Scheraga, this volume) can aid to avoid a single stress/single response approach. Such tools can help assure that people react on multiple fronts and respond in a more analytic way.

Implications for Institutional Design

The availability of multiple mental processes (analytic and experiential/affective) and the related multiple decision-making modes (calculation, affect, and recognition of rules, roles, or cases) make some processes and modes more suitable for addressing some problems rather than others. This has important implications for the design of institutions, as institutions can shape how decisions are being made, by making use of the different ways in which a problem is addressed (Engel and Weber 2007). A problem can be approached by seeking external or internal knowledge (expert advice), by conscious deliberation, or by spontaneous affective reactions. Each problem-solving mode generates different outputs. More deliberate reasoning can be achieved by raising the stakes. Similarly, introducing a justification requirement makes people aware of the complexity of the task at hand and makes accountability more salient, which in turn results in raised stakes. When immediate intervention is needed, it makes sense to match the current task with another preconfigured problem-solving mode that triggers emotions and an associated reaction (e.g., fight or flight behavior), for instance through recall from memory by priming.

Discount Rates

Discount rates tend to be lower when the outcome affects a client population rather than oneself. We may expect less discounting of the future for a social rather than an individual goal. Policy concerning investment in public goods (public health, mitigation of climate change) needs to distinguish between multiple self-regarding goals of individuals, social goals of families and friends, and civic goals.

The latest research on context-specific discount rates suggests that shorter-term environmental policy (whether the goal is to represent a constituency's general will or to structure citizens' behavior) can draw on the great body of research on discounting of financial outcomes and apply it to discounting of environmental outcomes, as long as care is taken to account for important contextual factors such as defaults, valence, and magnitude of the issue.

Risk Communication and Management Challenges

The chapters by Moser, Winkler et al., Easterling et al., and Mackey in this volume discuss how to improve risk communication by addressing (model) uncertainties more explicitly and by providing scenarios. In this chapter, we ask how people's experiential and affective processing and their aversion to uncertainty can be utilized constructively. While scenarios are often criticized, worst-case, best-case, most-likely case scenarios are a valuable tool that can help people plan for uncertainties, especially if they are sector-relevant (e.g., for agricultural decisions, city management, water resource management), spatially downscaled, and pertinent or relatable to the time horizons of decision makers.

As Winkler et al. demonstrate, if done well, as in the case of the Pileus Project, scenarios provide a good match to the non-probabilistic information processing of the experiential system. Scenario analysis is especially effective if it is presented in association with contingency plans, especially for worrisome worst- and bad-case scenarios. There are real and psychological benefits—the real benefits being an increased response speed and better responses; the psychological benefits being that perceived preparedness reduces anxiety.

CONCLUSIONS

The probabilistic nature of climate predictions for the Great Lakes poses both liabilities and opportunities. In the absence of clear action implications (which allow a feeling of control), awareness of climate risk may arouse too much anxiety—therein lies a liability. Uncertainty gets edited out, i.e., it is treated as being effectively zero, resulting in procrastination and decision avoidance. Furthermore, uncertainty can be used strategically to justify decisions, which are desired for other reasons (such as hidden agendas). On the positive side, uncertainty does not have to be an "enemy." The range of outcomes can serve as a natural impetus for contingency planning and thereby provide opportunities. If the development of forecast formats takes human information-processing modes and constraints into consideration, liabilities can be minimized and opportunities maximized. Consideration of the combination of analytic and experiential/affective processes can facilitate correct interpretation of climate forecasts and motivate forecast usage and adaptive risk-management actions.

Forecast formats and risk-management processes should be tailored to different segments of users. Users vary by the amount and sophistication of analytic processing, but time periods, incentives, and goals also differ across decision makers. For most users, it will pay to elicit optimal level of worry/concern. It would be useful to develop visualization tools to concretize the (temporally and spatially distant) impacts of climate change. The majority of decision makers/forecast users would benefit if statistical uncertainty measures were concretized. This could be achieved by the localization and downscaling of forecasts and predictions, the provision of analogies to previously experienced situations, and making explicit the distribution of different future cases (best case, most likely case, worst case, and likelihood of extreme events). Users of climate-variability forecasts and climate-change predictions or projections need to be provided with information about the degree of confidence, and an explanation of what those degrees of confidence signify.

Actions and choices can be influenced by strategic use of "framing." Situations can be described in ways that prime cross-group commonalities, social goals, and cooperation vs. differences, selfish goals, and competition. Depending on the desired response, communicators can choose reference points that depict alternatives as involving gains or losses. People tend to be risk seeking in the domain of losses, and risk averse for gains. For instance, protective or mitigative actions can be perceived as either involving costs and losses or benefits and opportunities. Both perceptions are true, but the attentional focus induced by the problem description often determines responses.

The design of effective risk-communication and risk-management processes must thus consider the above described constraints on human cognition and motivation, in addition to economic and institutional constraints. Knowledge about human capabilities and constraints provide useful tools. If ignored, many problems may seem more intractable than they are.

This chapter was prepared in part with funding from the National Science Foundation to the Center for Research on Environmental Decisions, Grant Number NSF-SES 0345840.

REFERENCES

Alloy, L.B., and L.Y. Abramson. 1982. Learned helplessness, depression, and the illusion of control. *Journal of Personality and Social Psychology* 42:1114–1126.

Alpert, M., and H. Raiffa, eds. 1969. A progress report on the training of probability assessors. In *Judgment under Uncertainty: Heuristics and Biases*, ed. by D. Kahneman, P. Slovic, and A. Tversky. Cambridge: Cambridge University Press.

Arora, P., N. Peterson, D.H. Krantz, D.J. Hardisty, and K. Reddy. 2012. To cooperate and not to cooperate: Using new methodologies and frameworks to understand how affiliation influences cooperation in the present and future. *Journal of Economic Psychology* (forthcoming).

Baron, J. 2000. Can we use human judgments to determine the discount rate? *Risk Analysis* 20:861–868.

Berbaum, K.S., E.A Franken Jr., D.D. Dorfman, E.M. Miller, R.T. Caldwell, D.M. Kuehn, and M.L. Berbaum. 1998. Role of faulty visual search in the satisfaction of search effect in chest radiography. *Academic Radiology* 5:9–19.

Bos, J.M., M.J. Postma, and L. Annemans. 2005. Discounting health effects in pharmacoeconomic evaluations: Current controversies. *Pharmacoeconomics* 23(7):639–649.

Brewer, M.B. 2001. The many faces of social identity: Implications for political psychology. *Political Psychology* 22(1):115–125.

Brewer, M.B., and R.M. Kramer. 1986. Choice behavior in social dilemmas: Effects of social identity, group size, and decision framing. *Journal of Personality and Social Psychology* 50(3):543–549.

Case, T., J. Fitness, D.R. Cairns, and R. Stevenson. 2004. Coping with uncertainty: Superstitious strategies and secondary control. *Journal of Applied Social Psychology* 34(4):848–871.

Chaiken, S., and Y. Trope. 1999. *Dual Process Theories in Social Psychology.* New York: Guilford Publications.

Chapman, G.B. 1996. Expectations and preferences for sequences of health and money. *Organizational Behavior and Human Decision Processes* 67:59–65.

———. 2002. Your money or your health: Time preferences and trading money for health. *Medical Decision Making* 20:410–416.

———. 2003. Time discounting of health outcomes. In *Time and Decision: Economic and Psychological Perspectives on Intertemporal Choice*, ed. by G.A. Loewenstein, D. Read, and R.F. Baumeister. New York: Russell Sage Foundation.

Damasio, A. 1995. *Descartes' Error: Emotion, Reason, and the Human Brain.* New York: Harper Collins.

Debreu, G. 1959. Topological methods in cardinal utility theory. In *Mathematical Methods in the Social Sciences*, ed. by K.J. Arrow, S. Karlin, P. Suppes. Stanford, CA: Stanford University Press.

Dudley, R.T. 1999. The effect of superstitious belief on performance following an unsolvable problem. *Personality and Individual Differences* 26(6):1057–1064.

Dunning, D., D.W. Griffin, J.H. Milojkovic, and L. Ross. 1990. The role of construal processes in overconfident predictions about the self and others. *Journal of Personality and Social Psychology* 59:1128–1139.

Engel, C., and E.U. Weber. 2007. The impact of institutions on the decision of how to decide. *Journal of Institutional Economics* 3:323–349.

Epstein, S. 1994. Integration of the cognitive and the psychodynamic unconscious. *American Psychologist* 49:709–724.

Felson, R.B., and G. Gmelch. 1979. Uncertainty and the use of magic. *Current Anthropology* 20:587–589.

Fischhoff, B., P. Slovic, S. Lichtenstein, S. Read, and B. Combs. 1978. How safe is safe enough? A psychometric study of attitudes towards technological risks and benefits. *Policy Sciences* 9:127–152.

Frederick, S., G. Loewenstein, and T. O'Donoghue. 2002. Time discounting and time preference: A critical review. *Journal of Economic Literature* 40(2):351–401.

———. 2003. Time discounting and time preference: A critical review. In *Time and Decision: Economic and Psychological Perspectives on Intertemporal Choice*, ed. by G.F. Loewenstein, D. Read, and R. Baumeister. New York: Sage.

Gattig, A., and L. Hendrickx. 2007. Judgmental discounting and environmental risk perception: Dimensional similarities, domain differences, and implications for sustainability. *Journal of Social Issues* 63(1):21–39.

Guyse, J.L., L.R. Keller, and T. Eppel. 2002. Valuing environmental outcomes: Preferences for constant or improving sequences. *Organizational Behavior and Human Decision Processes* 87:253–77.

Hansen, J., S.M. Marx, and E.U. Weber. 2004. The role of climate perceptions, expectations, and forecasts in farmer decision making: The Argentine Pampas and South Florida. In *IRI Technical Report 04–01.* Palisades, NY: International Research Institute for Climate Prediction.

Hardin, G. 1968. The tragedy of the commons. *Science* 162(3859):1243–1248.

Hardisty, D., and E.U. Weber. 2009. Discounting future green: Money vs. the environment. *Journal of Experimental Psychology: General* 138(3):329–340.

Haselton, M.G., and D. Nettle. 2006. The paranoid optimist: An integrative evolutionary model of cognitive biases. *Personality and Social Psychology Review* 10(1):47–66.

Hertwig, R., G. Barron, E.U. Weber, and I. Erev. 2004. Decisions from experience and the effect of rare events. *Psychological Science* 15:534–539.

Hogarth, R.M. 1987. *Judgement and Choice.* 2nd ed. New York: Wiley.

IPCC. 2001a. *IPCC Special Report on Emission Scenarios.* Intergovernmental Panel on Climate Change.

———. 2001b. *IPCC Third Assessment Report: Climate Change 2001.* Intergovernmental Panel on Climate Change.

———. 2007. *IPCC Fourth Assessment Report: Climate Change 2007 Synthesis Report.* Intergovernmental Panel on Climate Change.

Keeney, R.L., and H. Raiffa. 1976. *Decisions with Multiple Objectives: Preferences and Value Tradeoffs.* New York: John Wiley.

Keinan, G. 1994. Effects of stress and tolerance of ambiguity on magical thinking. *Journal of Personality and Social Psychology* 67:48–55.

———. 2002. The effects of stress and desire for control on superstitious behavior. *Personality and Social Psychology Bulletin* 28:102–108.

Keren, G. 1987. Facing uncertainty in the game of bridge: A calibration study. *Organizational Behavior and Human Decision Processes* 39:98–114.

Kiecolt-Glaser, J., P.T. Marucha, W. Malarkey, A. Mercado, and R. Glaser. 1995. Slowing of wound healing by psychological stress. *Lancet* 346:1194–1196.

Koriat, A., S. Lichtenstein, and B. Fischhoff. 1980. Reasons for confidence. *Journal of Experimental Psychology: Human Learning and Memory* 6:107–118.

Krantz, D.H., and H.C. Kunreuther. 2007. Goals and plans in decision making. *Judgment and Decision Making* 2(3):137–168.

Krantz, D.H., R.D. Luce, P. Suppes, and A. Tversky. 1971. *Foundations of Measurement.* Vol. 1, *Additive and Polynomial Representations.* New York: Academic Press.

Langer, E.J. 1975. Illusion of control. *Journal of Personality and Social Psychology* 32(2):311–328.

Leiserowitz, A. 2004. Surveying the impact of "The Day After Tomorrow." *Environment* 46(9):23–44.

———. 2006. Climate change risk perception and policy preferences: The role of affect, imagery, and values. *Climatic Change* 77(1–2):45–72.

Letson, D., C.E. Laciana, F. Bert et al. 2009. The value of perfect ENSO phase predictions for agricultural production: Evaluating the impact of land tenure and decision objectives. *Climatic Change* 97(1–2):14–170.

Lichtenstein, S., and B. Fischhoff. 1977. Do those who know more also know more about how much they know? The calibration of probability judgments. *Organizational Behavior and Human Decision Processes* 65:117–137.

Lichtenstein, S., B. Fischhoff, and L. Phillips. 1982. Calibration of probabilities: The state of the art. In *Judgment under Uncertainty: Heuristics and Biases*, ed. by D. Kahneman, P. Slovic, and A. Tversky. Cambridge: Cambridge University Press.

Lichtenstein, S., and P. Slovic. 2006. *The Construction of Preference*. Cambridge: Cambridge University Press.

Linville, P.W., and G.W. Fischer. 1991. Preferences for separating and combining events: A social application of prospect theory and the mental accounting model. *Journal of Personality and Social Psychology Bulletin* 60:5–23.

Loewenstein, G. 1988. Frames of mind in intertemporal choice. *Management Science* 34:200–214.

Loewenstein, G.F., and D. Prelec. 1993. Preferences for sequences of outcomes. *Psychological Review* 100:91–108.

Loewenstein, G.F., E.U. Weber, C.K. Hsee, and N. Welch. 2001. Risk as feelings. *Psychological Bulletin* 127(2):267–286.

Marx, S.M., E.U. Weber, B.S. Orlove et al. 2007. Communication and mental processes: Experiential and analytic processing of uncertain climate information. *Global Environmental Change* 17:47–58.

Maslow, A.H. 1943. A theory of human motivation. *Psychological Review* 50:370–396.

Mischel, W., J. Grusec, and J.C. Masters. 1969. Effects of expected delay time on the subjective value of rewards and punishments. *Journal of Personality and Social Psychology* 11:363–373.

Murphy, A.H., and B.G. Brown. 1984. Comparative evaluation of objective and subjective weather forecasts in the United States. *Journal of Forecasting* 3:369–393.

Murphy, A.H., and R.L. Winkler. 1984. Probability Forecasting in Meteorology. *Journal of the American Statistical Association* 79(3):489–500.

Ostrom, E., T. Dietz, N. Dolsak, P.C. Stern, S. Stonich, and E.U. Weber, eds. 2002. *The Drama of the Commons*. Washington DC: National Academy Press.

Payne, J.W., J.R. Bettman, and E.J. Johnson. 1993. *The Adaptive Decision Maker*. Cambridge: Cambridge University Press.

Read, D. 2001. Is time-discounting hyperbolic or sub-additive? *Journal of Risk and Uncertainty* 23:5–32.

Russo, J.E., and P.J.H. Schoemaker. 1992. Managing overconfidence. *Sloan Management Review* 33(2):7–17.

SAP/USP. 2009. *Global Climate Change Impacts in the United States*. U.S. Global Change Research Program.

Sloman, S.A. 1996. The empirical case for two systems of reasoning. *Psychological Bulletin* 1(119):3–22.

Slovic, P. 1995. The construction of preference. *American Psychologist* 50:364–371.

————. 1997. Trust, emotion, sex, politics, and science: Surveying the risk-assessment battlefield. In *Psychological Perspectives to Environmental and Ethical Issues in Management*, ed. by M. Bazerman, D. Messick, A. Tenbrunsel, and K. Wade-Benzoni. San Francisco: Jossey-Bass.

Slovic, P., M. Finucane, E. Peters, and D.G. MacGregor. 2002. The affect heuristic. In *Intuitive Judgment: Heuristics and Biases*, ed. by T. Gilovich, D. Griffin, and D. Kahneman. New York: Cambridge University Press.

————. 2007. The affect heuristic. *European Journal of Operational Research* 177(3):1333–1352.

Slovic, P., B. Fischhoff, and S. Lichtenstein. 1980. Facts and fears: Understanding perceived risk. In *Societal Risk Assessment: How Safe Is Safe Enough?*, ed. by R. Schwing and J.W.A. Albers. New York: Plenum Press.

————. 1986. The psychometric study of risk perception. In *Risk Evaluation and Management*, ed. by V.T. Covello, J. Menkes, and J. Mumpower. New York: Plenum Press.

Sunstein, C.R. 2006. Precautions against what? The availability heuristic, global warming, and cross-cultural risk perceptions. *Climatic Change* 77(1–2):195–210.

Thaler, R.H. 1981. Some empirical evidence on dynamic inconsistency. *Economics Letters* 8:201–207.

Thompson, S.C. 1981. Will it hurt less if I can control it? A complex answer to a simple question. *Psychological Bulletin* 90: 89–101.

Treadwell, J.R. 1997. Discounting and independence in preferences between health sequences. Ph.D. dissertation, Department of Psychology, University of Washington.

Tversky, A., and D. Kahneman. 1973. *Judgment under Uncertainty, Utility, Probability, and Decision Making*. Edited by G. Wendt and P. Vlek. Boston: D. Reidel Publishers.

————. 1979. Prospect theory: An analysis of decision under risk. *Econometrica* 47:263–292.

————. 1992. Advances in prospect theory: Cumulative representation of uncertainty. *Journal of Risk and Uncertainty* 5(4):297–323.

Vallone, R., D.W. Griffin, S. Lin, and L. Ross. 1990. The overconfident prediction of future actions and outcomes for self and other. *Journal of Personality and Social Psychology* 85:582–592.

Wager, T.D., J.K. Rilling, E.E. Smith et al. 2004. Placebo-induced changes in fMRI in the anticipation and experience of pain. *Science* 303(5661):1162–1167.

Wager, T.D., D.J. Scott, and J.K. Zubieta. 2007. Placebo effects on human mu-opioid activity during pain. *Proceedings of the National Academy of Sciences* 104:11056–11061.

Weber, B.J., and G.B. Chapman. 2005. The combined effects of risk and time on choice: Does uncertainty eliminate the immediacy effect? Does delay eliminate the certainty effect? *Organizational Behavior and Human Decision Processes* 96(2):104–118.

Weber, E.U. 1997. Perception and expectation of climate change: Precondition for economic and technological adaptation. In *Psychological Perspectives to Environmental and Ethical Issues in Management*, ed. by M. Bazerman, D. Messick, A. Tenbrunsel and K. Wade-Benzoni. San Francisco: Jossey-Bass.

———. 2006. Experience-based and description-based perceptions of long-term risk: Why global warming does not scare us (yet). *Climatic Change* 70:103–120.

———. 2011. Achieving sustainable development: Behavior change through goal priming and judicious decision mode selection. In *Is Sustainable Development Feasible?*, ed. by J.S. Sachs and P. Schlosser. New York: Columbia University Press (in press).

Weber, E.U, and E.J. Johnson. 2006. Constructing preferences from memory. In *The Construction of Preference*, ed. by S. Lichtenstein and P. Slovic. New York: Cambridge University Press.

Weber, E.U., E.J. Johnson, K. Milch, H. Chang, J. Brodscholl, and D. Goldstein. 2007. Asymmetric discounting in intertemporal choice: A query theory account. *Psychological Science* 18:516–523.

Weber, E.U., and P.G. Lindemann. 2007. From intuition to analysis: Making decisions with your head, your heart, or by the book. In *Intuition in Judgement and Decision Making*, ed. by H. Plessner, C. Betsch, and B.T. Betch. Mahwah, NJ: Lawrence Erlenbaum.

Weber, E.U., S. Shafir, and A.R. Blais. 2004. Predicting risk sensitivity in humans and lower animals: Risk as variance or coefficient of variation. *Psychological Review* 111:430–445.

Agricultural Adaptation to Climate Change

Is Uncertain Information Usable Knowledge?

WILLIAM E. EASTERLING, CLARK SEIPT, ADAM TERANDO, AND XIANZENG NIU

THE EARTH LIKELY IS COMMITTED TO AT LEAST 0.5° TO 0.6°C OF future warming in response to historical atmospheric accumulations of greenhouse gas emissions, regardless of steps taken, if any, to mitigate future emissions (Karl and Trenberth 2003; IPCC 2007). Unabated future emissions will surely add even more warming and climate changes. The effects of future climate change, mitigated or not, on critical ecosystem services such as food production are predicted to require adaptation in order to avoid or minimize losses or seize gains (NRC 2001; Gitay et al. 2001).

Adaptation to climate change is defined here as adjustment to or by ecosystems and society in response to climate stimuli and their impacts (Smit et al. 2000). Bradshaw et al. (this volume) posit that adaptation can take several forms along the lines of such attributes as proactive versus reactive (e.g., manipulation of trade policy versus change in existing on-farm tillage systems), short versus long time scale (e.g., use of seasonal climate predictions versus investment in fundamental adaptive research), local versus national scale (e.g., farm-level soil and water conservation practices versus federal crop insurance policy), and technological versus institutional (e.g., implementation of irrigation infrastructure versus modification of groundwater management policy). The implicit goal of adaptation is to avoid damage to ecological and social form and function so as to maintain current ecological integrity and related social welfare. Although some researchers use the terms *mitigation* and *adaptation* interchangeably to collectively identify actions to ameliorate climate change and the consequences thereof, they are kept distinct here where the focus is exclusively on adaptation as defined above.

No matter what form adaptation takes, it will require a sustained stream of reliable information from the scientific community. That is, it will require *usable knowledge* in order to remain effective. Usable knowledge is accurate information that is useful to decision makers (Haas 2004). Haas argues that it has a number of qualifying traits: (1) it is tractable to its users—i.e., it can be practically applied; (2) it is credible—i.e., it is believed to be true; (3) it is legitimate—i.e., it is thought to derive from a valid scientific process; and (4) it is salient—i.e., its arrival is timely with respect to important policy or management decisions.

One of the formidable challenges to the efficient exchange of usable knowledge about climate change adaptation between the research community and the managers, policymakers, and other stakeholders who would benefit from it is the large inherent uncertainty in that knowledge. Climate change is an uncertain science. Although great progress has been made toward pinpointing the sensitivity of the climate system to rising greenhouse gas concentrations, it is still expressed by scientists as a sizable range. But does that mean that uncertain information about climate change and potential adaptation strategies cannot, ipso facto, become usable knowledge? We assert that the answer is an emphatic no. It is not reasonable to hold all scientific knowledge to a standard of indisputable fact in order for it to be considered usable knowledge. Were that not the case, then neither weather forecasts nor projected hurricane paths would graduate to become usable knowledge. However, it does follow that the scientific community has an obligation to make uncertainties explicit as a necessary condition for its findings to be transformed into usable knowledge, and to be attentive to issues of how uncertainty is processed in decision making, as reviewed by Marx and Weber (this volume) and Moser (this volume).

The purpose of this chapter is to identify and characterize the major sources of uncertainty concerning the estimation of the effects of climate change on food security and the prospects for effective adaptation. For convenience, the sources of uncertainty are classified into three categories: *fundamental, structural,* and *parametric.* We include two case studies of research designed to illustrate the importance of unpacking the effect of scientific uncertainty on the adaptive process itself: the first examines the effect of path dependence on adaptation of crop production to climate change, and the second suggests a beginning discourse between scientists and farmers over the utility of climate forecasts. We consider uncertainties across the spectrum of biophysical and socioeconomic (and policy) knowledge needed for effective adaptation. The goal is to suggest ways to illuminate uncertainty in an effort to promote adaptive research as usable knowledge. We note that this is of particular importance to the Great Lakes region because agriculture is the second or third largest sector of the economy of most Great Lakes states and provinces, and that sector is strongly affected by global trade, and thus global, not just local, changes in agriculture as a result of climate change (see also Winkler, this volume).

A NOTE ON UNCERTAINTY AND IMPACT ASSESSMENT

Estimates of the consequences of climate change for ecosystems and society are the least certain links in the chain of information flow that begins with the predicted sensitivity of the climate system to increased radiative forcing and ends with predicted consequences for ecosystems and society. State-of-the-science impact assessment modeling assimilates all of the essential

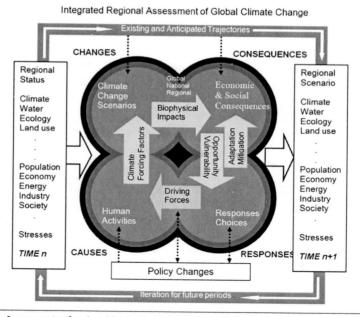

Figure 1. Components of regional integrated impact assessment modeling
Source: Penn State Center for Integrated Regional Analysis.

components of the integrated biosphere-social-economic system in order to quantify the interactions of climate change, biophysical and socioeconomic form and process, adaptation, and mitigation actions (figure 1). Not only does the uncertainty of climate change scenarios propagate from one component to the next, but new uncertainty is introduced with each additional component of the impact assessment (figure 2). For example, in one assessment of the uncertainty of future predictions of climate change impacts, Parry et al. (2005) found that the variation of predicted food production around the global average late in the twenty-first century is more than 50 percent of the mean climate change signal. That uncertainty is compounded by the uncertainty of how geopolitics and technology development, for example, will influence food production. This high level of compounded uncertainty weakens the linkage between climate change and food production for policy purposes. The challenge of impact assessment is to convey a strong sense to the stakeholders/decision makers of the likelihood of a given impact occurring or not occurring, or the successful avoidance of impacts by adaptation (NRC 2003).

BACKGROUND

One of the remarkable human achievements of the twentieth and early twenty-first centuries was the success of the world's farmers in increasing global agricultural food production faster than the growth in global food demand. Population and per capita income are the two major drivers of

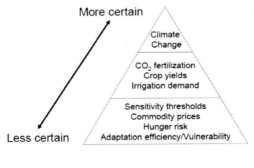

Cascading Uncertainty

Figure 2. Cascading uncertainty in the chain of climate change knowledge

global food demand. Between 1960 and 2000, the population of the earth doubled. Globally averaged per capita income in the 1990s rose at an annual rate of 1.2 percent. However, the prices for the major cereal crops (rice, wheat, and maize), which account for about two-thirds of the calories we consume directly, or indirectly through livestock, fell by 60 percent through the 1990s when adjusted for inflation (Bruinsma, 2003).

The sequence of events was that global food production grew at a rate of 2.2 percent per year (p.a.) over the period 1969 to 1999, driven by a global explosion of new technologies such as high-yielding crop varieties and increasingly effective chemical inputs such as pesticides and nitrogen fertilizers, thus outstripping growth in demand. Provisional estimates by FAO (2005) project annual production growth rates of 1.6 percent out to 2015 and 1.3 percent out to 2015–2030. This is in perfect step with anticipated slowing growth in demand due to continued deceleration of population growth and rising nutritional levels among many of the world's poorest people.

The world's food supply is generally predicted to be in extraordinarily good shape over the course of the twenty-first century, provided that climate change and other major environmental stresses do not slow the growth in world agricultural capacity to below the growth in demand. There is the rub. The preponderance of global agricultural studies (e.g., Adams et al. 1999; Fischer et al 2001; Parry et al 1999) have established, although incompletely, that climate change is not likely to diminish global agricultural capacity by more than a few percent, if at all, by 2050, when taking into account regions that may benefit (i.e., North America, Europe) and regions that may suffer (i.e., the Tropics). Any losses would be on top of substantial gains in world output due to technological improvements and wider distribution of science-based production—global agricultural capacity, climate change apart, is projected to be about 55 percent greater than current by 2030. A small but growing suite of modeling studies generally predict that world grain (real) prices are likely to continue to decline through the first 2° to 3°C of warming. However, warming that exceeds 2 to 3 degrees puts the world into uncharted territory where grain prices are expected to reverse direction and begin to rise—hence, *2 to 3 degrees of warming appears to be a crucial threshold or tipping point for crop prices* (Easterling et al. 2007). Since some recent analyses suggest that the 2-degree threshold may be hard to obtain given current policy trajectories (Fawcett et al. 2009), the effects of passing the threshold need serious consideration; but most analyses focus on a world in which we stay below 2 to 3 degrees of warming.

While the global food situation looks manageable for the first half of the twenty-first century under a 2 to 3 degree warming scenario, there are reasons for concern at regional levels. The Third Assessment Report of the Intergovernmental Panel on Climate Change (IPCC), summarized by Easterling and Apps (2005), reported that a number of models simulate the capacity of temperate crops (wheat, maize, rice) to absorb 2 to 3 degrees of warming before showing signs of stress. The IPCC Fourth Assessment Report (Easterling et al. 2007) found that agronomic adaptation extends the threshold to beyond 5°C of warming for those crops. Tropical crops exhibit immediate yield decline with even the slightest warming because they are currently grown under conditions close to maximum temperature tolerances—even a little warming sends them lower than current levels. Adaptation gives tropical regions a buffer of approximately 3°C of warming before yields of those crops dip below current levels. In spite of this slight buffer, the news is not good for developing countries in the Tropics, especially Africa. After about 4° to 5°C of warming, the news is not good anywhere.

Two regions that are likely to experience large negative impacts of climate change on agricultural production are Asia and Africa. According to the IPCC-AR4, studies indicate that rice production across Asia could decline by nearly 4 percent over the twenty-first century. In India, a 2°C increase in mean air temperature could decrease rice yield by about 0.75 metric tons per hectare and cause a decline in rain-fed rice in China by 5 to 12 percent. Sub-Saharan Africa could lose a substantial amount of cropland due to climate-change-induced land degradation. Based on results generated with the HadCM3 scenarios, which are scenarios derived from the projected future climate change by a coupled atmosphere-ocean general circulation model developed at the UK Hadley Centre for Climate Prediction and Research (http://www.ipcc-data.org/sres/hadcm3_download .html), and the future socioeconomic development paths, as many as 40 food-insecure countries of sub-Saharan Africa, with a projected total population in 2080 of approximately 1 to 3 billion, may lose on average 10 to 20 percent of their cereal-production potential caused by climate change.

In addition, the rise in the use of biofuels could further increase the food insecurity in the world's developing countries. Biofuels are considered as renewable and clean energy that can be used to replace fossil fuels to mitigate CO_2 emissions. However, the main sources of biofuels are first-generation biofuel crops, such as corn, soybeans, sugar cane, and palm oil that have traditionally been used for food production. Expansion in biofuel production will inevitably impose serious impacts on world food production, hence on food prices and food security, including increasing competition about the use of farmland and water resources (Escobar et al. 2009). The causes for recent-year food price increases are related, at least partially, to the increase in the oil prices and the growing use of biofuels produced out of corn and sugar cane (Escobar et al. 2009).

SOURCES OF UNCERTAINTY ABOUT AGRICULTURAL IMPACTS AND ADAPTATION

Broadly speaking, the uncertainty that surrounds estimates of the consequences of climate change for ecosystem services and society derives from researchers' overreliance on "point estimate" methodologies wherein estimates are the product of a series of "best guess" assumptions (Moss and Schneider 2000; Roughgarden and Schneider 1999). That is, the estimates are robust only as long as all boundary assumptions—e.g., climate change characteristics, system sensitivity to climate change, adaptive potential, socioeconomic context—are satisfied. Such would be a long shot at best. We assert that the point-estimate case is symptomatic of much deeper and more nuanced sources of uncertainty, especially with respect to a complex human-environment system such as agricultural production. There are endless ways to classify uncertainty, but major sources of uncertainty concerning scientific information about climate change and its consequences have been usefully classified into three categories (Manning et al. 2004): 1) *fundamental uncertainty*; 2) *structural uncertainty*; and 3) *value or parametric uncertainty*. Those categories provide a framework used here for reviewing sources of uncertainty vis-à-vis agricultural impacts and adaptation.

Fundamental Uncertainty

Future predictions are always subject to surprises that occur when perceived reality departs radically from expectations (Kates and Clark 1996). Tom Schelling illustrates this by pointing out the revolutionary surprises a futurist writing at the turn of the twentieth century would have been forced to explain a few short years later, including the advent of self-propelled flight, mechanized agriculture, anesthesia and antibiotics, electric light, the transistor radio, the telephone, the great flu epidemic, the Bolshevik Revolution, and World War I, to name just a few (NRC 1983). Those surprises completely reframed the social, political, economic, and technological fabric of the early twentieth

Type	Indicative examples of sources	Typical approaches and considerations
Fundamental uncertainty	Situations so novel that no existing model applies or systems that are so non-linear or chaotic that modeling is intractable. This includes complex systems that, on occasion, produce emergent properties. Examples include long-range prediction of future technology or the evolution of social, economic, and political systems.	Use of scenarios spanning a plausible range, ranges from ensembles of model runs.
Structural uncertainty	Models with incomplete or competing conceptual frameworks, lack of agreement on model structure, ambiguous system boundaries or definitions, significant processes or relationships wrongly specified or not considered.	Specify assumptions and system definitions clearly, compare models with observations for a range of conditions, assess maturity of the underlying science and degree to which understanding is based on fundamental concepts tested in other areas.
Value or Parametric uncertainty	Models with inaccurate specifications of known processes, which may be attributed to poor parameterizations. Such cases often arise from reliance on partial or incomplete observation.	Test model parameter sensitivity to data subsets. Compare the same parameterizations from different models.

Figure 3. Categories of uncertainty
Source: Adapted from Manning et al. (2004).

century. The point here is that we should expect that unpredictable events will alter how the world will co-evolve with climate change.

Unpredictable surprise is the result of *fundamental uncertainty*, defined as a system state for which no model exists, or its boundary conditions are too complex or chaotic to be satisfied by a predictive model. Fundamental uncertainty is thus the least tractable form of uncertainty with respect to delivering usable knowledge to decision makers. Examples of fundamentally uncertain, but plausible, futures that would alter the scope of agricultural adaptation to climate change include:

Major changes in food preferences. A large-scale change in peoples' dietary preferences, such as a radical global transition away from meat toward higher direct consumption of high-protein legumes and grains, would trigger a chain reaction beginning with sweeping changes in consumption patterns and other market behaviors and reaching all the way to production decisions at the farm gate. Preposterous (to many agriculturalists) as this may seem, one could easily imagine several valid reasons why this might occur, including concerns over health risks and rising energy costs. But ultimately, such a radical transition would be the outcome of a complex set of human choices for which there is no predictive model.

Sudden arrival of new pesticide-resistant invasive species. The invasion of new pest species (insects and weeds) and pathogens, perhaps aided by climate change or by evolution to new genomes, is a potential threat to agriculture everywhere (see Hall and Root, this volume). The broad scientific principles governing "invasive potential" for all pests in all environmental circumstances are not known (NRC 2002). The lack of coordinated observing has hindered the derivation of such broad principles. The modes of transmission of these pests and pathogens are numerous and complex, but most boil down to some form of human activity (e.g., food shipments, inadvertent importation, deliberate introduction), although weather and climate events (e.g., hurricanes and other mobile weather systems with wind-transport potential) are also important. The adaptive process itself could be a source of invasive species, especially adaptations that involve translocation of crop species from other regions. The many ways that humans facilitate the spread of invasive species is an almost intractable modeling challenge and is a primary source of fundamental uncertainty.

Global or regional conflict, including bioterrorism, that alters world food trade relations, damages agricultural production systems, or potentially confiscates them from markets. Wars and other forms of armed conflict disrupt food production and food supply chains, and are known to be strong determinants of food shortage and famines (FAO 2003).

Increasingly, biosecurity threats to food-production systems are becoming a possibility. The plausibility of war, in combination with natural disasters such as tsunamis and droughts, causing food crises has been demonstrated repeatedly in recent decades (Pingali et al. 2005). The Ethiopian famines of the late 1970s and early 1980s will stand forever as a marked testimonial of this lethal combination. The likelihood that such events will occur again is high, but there is little basis for predicting their onset, severity, duration, or location.

Collapse in energy markets; rapid onset of energy insecurity. The International Energy Agency projects global demand for primary energy resources in 2030 to increase by 60 percent over current demand. Global annual consumption of petroleum in 2000 was 3 percent of proven reserves; coal was 0.5 percent of proven reserves; and natural gas consumption was 1.6 percent of proven reserves (Chow et al. 2003). While economic scarcity of these important resources is not imminent, it is within sight. A long-term increase in energy prices would present a major challenge to agricultural adaptation because virtually every part of the food production and distribution system is highly dependent on cheap and reliable energy (fertilizers, irrigation, tillage and harvesting, processing, transport to markets), even in developing countries. The plausibility of a major global energy crisis is very real, but fundamentally uncertain and itself subject to uncertainty over conflict and other forms of human-induced and environmental disruptions.

All of the above examples are equally plausible future developments that could individually or in combinations alter the type, timing, and intensity of adaptation to climate change across a range of scales and geographic locations. Individually, they represent singularities that are easy to imagine, but difficult, perhaps impossible, to predict.

One way to constrain fundamental uncertainty is through scenario analysis in which a range of future scenarios is specified that bounds the realm of plausible but fundamentally uncertain futures. The development and distribution of the SRES scenarios by the IPCC (IPCC 2000; SAP 2008) was an important step in that direction. These scenarios consist of alternative global greenhouse-gas-emissions trajectories for the twenty-first century based on different assumptions concerning future pathways of economic growth, globalization, and attitudes toward the basic concept of sustainability—these are the major determinants of greenhouse gas emissions. The alternatives are meant to encompass a large range of plausible futures, therefore bounding future greenhouse gas trajectories.

Structural Uncertainty

Why do general circulation models disagree on the regional details of predicted future climate change? Why do the yield predictions of physiologically explicit crop-growth models agree in some locations and disagree in other locations when run with identical climate and other input data? There is no single answer to these questions, but a trait common to both of these and similar classes of questions is the existence of large structural uncertainties among methodological approaches. *Structural uncertainty* results when models have alternative conceptual frameworks, divergent sets of state- and path-dependent variables, or missing or wrongly specified variable-process relations. The types of structural uncertainty that are especially challenging to the production of usable knowledge of agricultural adaptive potential include, for example, (1) competing explanations of the magnitude of CO_2 direct effects on crop growth and yield; (2) lack of physiological detail concerning the effects of plant pests and pathogens on yields; (3) lack of physiological and integrated assessment models for estimating adaptive potential for livestock production; (4) limited capacity to account for the synergistic effects of multiple stresses and climate change on agricultural production systems; and (5) limited scope of model-relevant adaptation.

Role of atmospheric CO_2 effects. The degree to which increasing atmospheric CO_2 concentrations will enhance photosynthesis and water-use efficiency in certain types of crops, thus offsetting climate change impact, is not well established. Part of the problem is the proliferation of experimental approaches that have been used to quantify this enhancement. A recent review of open-field Free-Air Carbon Dioxide Enrichment (FACE) studies concluded that the positive effects of elevated CO_2 on several food crops are substantially less than those of previous experiments conducted in relatively more controlled enclosures and chamber environments (Long et al. 2006). FACE is argued to be the most realistic experimental design yet, vis-à-vis real farming environments. Because of the differences between FACE and older experimental results, Long et al. have argued that crop models—calibrated with the older data—may simulate CO_2 sensitivity too strongly, thus offsetting too much of the potentially negative effects of climate change on crop yields. Hence, estimates of world food supply under climate change, as reviewed by Gitay et al. (2001), are called into question as being overly optimistic and in need of downward revisions (Long et al. 2006). Tubiello et al. (2007) dispute those assertions by arguing that Long's analysis is an artifact of data-pooling methods that accentuate differences; the much smaller differences that exist between FACE and earlier data are of little consequence for crop modeling, hence world food supply projections. This is an important illustration of the case of uncertainty rising rather than being reduced by additional research results.

Lack of physiological detail concerning the effects of plant pests and pathogens on yields. World application of pesticides in 2000 was 3.75 million metric tons and is projected to rise to 6.55 million metric tons by 2020

(Tillman et al. 2001). Climate change likely will change the timing, species, and intensities of pest and pathogen outbreaks in major agricultural areas (Gitay et al. 2001). Only modest progress has been made in understanding pest and pathogen response to climate change, although pre-harvest losses of food and cash crops to pests is estimated at 42 percent of global production potential (Gitay et al. 2001). Particular uncertainties persist in understanding competition between C_3 and C_4 species under higher CO_2 when one species is a weedy pest and the other is a food or feed species (Ziska 2003). Also, little is known about the interactions between increasing atmospheric CO_2 and rising temperature. The paucity of usable knowledge about climate change effects on all aspects of pests and pathogens seriously limits understanding of how best to adapt to those effects.

Lack of physiological models of livestock response to climate change. Approximately 16 percent of the calories consumed by the world's population comes from animals. Meat consumption (beef, pork, poultry) in developed countries is projected to grow at an annual rate of 0.5 percent in developed countries and nearly 3 percent in developing countries to the year 2020 (Delgado et al. 1999). Knowledge of the effects of climate change on livestock animals relies mainly on experimental studies. Physiologically based simulation modeling appropriate for predicting climate change response of livestock has only recently begun. Frank et al. (2001) predict declines in U.S.-confined swine, beef, and dairy milk production of 1.2 percent, 2 percent, and 2.2 percent, respectively, by 2050 in response to a climate change simulation by the Canadian Global Climate Change program. The lack of robust models with which to test the effectiveness of simple livestock adaptation strategies means that a significant link in the food chain is missing from the knowledge base.

Accounting for the synergistic effects of multiple stresses and climate change on agricultural production systems. Most of our insights about how agricultural systems may respond to environmental change have been obtained by research approaches that focus on individual stresses (e.g., air pollution, climate variability and change, pests and pathogens, loss of genetic diversity, desertification and land degradation) most often in isolation from one another (Easterling 2006). Interactions among stressors rarely are explicit, even within integrated assessment modeling frameworks. Thus, assumptions that all else is equal have applied to model simulations of adaptive agricultural strategies in response to climate change. Gitay et al. (2001) found only a few studies that considered interactions among stressors. For example, research on the interacting effects of increased atmospheric CO_2 and O_3 levels simultaneously found that the CO_2 compensated for the O_3 damage

Global

Climate change

Widening agricultural trade deficits

Local and Regional

Rapid population growth

Emerging infectious diseases, i.e., HIV/AIDS

Pest outbreaks

Lack of safe drinking water, public health

Rising incomes leading to dietary changes, i.e., more meat consumption

Political instability and civil strife

Long-term drought

Land degradation

Figure 4. Stressors on sub-Saharan Africa food security
Source: Easterling (2006).

to wheat biomass production, but not to final economic yield. Not all of the stressors that are likely to affect agricultural adaptation to climate change are environmental. Many have social origins. Figure 4 lists an incomplete compilation of stressors to agricultural production in sub-Saharan Africa. No research has studied them in totality, although it is virtually certain that their interactions with climate change impacts are substantial and may even supersede the individual effect of climate change on any one part of the food production system. Without a modeling framework that permits sensitivity analysis of the interactive effects of climate change and social and environmental stressors, there is unconstrained structural uncertainty concerning the net effectiveness of adapting to one stressor (climate change) while everything else is held constant.

Small scope of model-relevant adaptation strategies. There is a multitude of potential agricultural adaptation strategies for dealing with climate change; the full portfolio of these strategies has been reviewed by a number of researchers (Easterling 1996; Easterling et al. 2005; Smit et al. 2000). However, it is possible to simulate only a small subset of those strategies in contemporary simulation modeling schemes (figure 5). There have been limited attempts to simulate the full suite of adaptation possibilities of farmers by manipulating historical relationships between land rent (economic returns from land ownership) and average climate conditions in initial econometric models (e.g., Mendelsohn et al. 1994; Mendelsohn

Autonomous Adaptation	*Simulation Potential with Existing Models*	*Planned Adaptation*	*Simulation Potential with Existing Models*
Timing of operations such as planting and harvesting	High	Improved climate monitoring and outreach	Low
Livestock or crop species or cultivar choices	High	Research and strategic planning to facilitate adaptation	Low
Irrigation strategies	High	Government subsidies to assist farmers and related businesses in transition to new production strategies	Low
Tillage strategies	High	New infrastructure, policies and institutions to support new management and land use arrangements	Low
Water harvesting strategies	Low	Enhanced investment in irrigation infrastructure and efficient water use technologies	Low
Enterprise diversification	Low	Revising land tenure arrangements including attention to well-defined property rights	Low
Pest management strategies	Low	Establishment of accessible, efficient markets for products and inputs (seed, fertilizer, labor etc.) and for financial services including insurance	Low

Figure 5. Ease of simulation of autonomous and planned agricultural adaptations

Autonomous Adaptation	Simulation Potential with Existing Models	Planned Adaptation	Simulation Potential with Existing Models
Use of climate information such as forecasts	Moderate	Ensuring appropriate transport and storage infrastructure	Low
Market-driven adjustments to resource allocations (i.e., land use) and trade	Moderate	Application of adaptive management principles	Low

Figure 5. Ease of simulation of autonomous and planned agricultural adaptations (*continued*)

et al. 2004); this approach presumes that land rent implicitly embeds all actions by farmers to adapt to their climatic resources. But severe boundary assumptions, especially stationarity of model parameters, restrict the applicability of Ricardian predictions to future climate change situations. This means that most of the knowledge concerning the potential effectiveness of agricultural adaptation to climate change is based on modeling schemes that embed a small sample of the hypothetically important adaptation possibilities. From the standpoint of uncertainty, this is a case of incomplete modeling structures.

The rudimentary state of knowledge concerning the underlying *process* of adaptation is a major source of structural uncertainty. By this we mean the uncertainty introduced by the lack of formal process-based structures in simulation models that can predict the effectiveness of alternative adaptation strategies within the overriding social and economic conditions that warrant the use of those strategies. In most instances where adaptation is explicitly modeled, an ad hoc scheme for introducing and testing various adaptive strategies was used. In such schemes, farmers are implicitly assumed to react instantly and flawlessly to climate change. They are, in short, perpetual optimizers. In reality, adaptation is not nearly as optimal or effective as it is in a simulation model.

Several researchers have approached the problem of representing adaptation as a dynamic process by using the induced innovation process (Hyami and Ruttan 1985; Ruttan 1997) as a theoretical framework. This approach has much promise as it attempts to make the adaptation process endogenous to farm-level production functions. It posits that innovation is induced by rising prices of critical production inputs for which there are no readily available substitutes. Although climate is an unpriced factor of production, some analysts have developed schemes for incorporating climate variability into the induced innovation hypothesis framework (Smithers and Blay-Palmer 2001; Chhetri and Easterling, 2008). A related unsolved problem in the effective simulation of agricultural adaptation is how to deal with the inefficiencies brought about by path dependence of farmers and their supporting institutions in how they perceive climate change, and then have the flexibility to reflexively adapt.

Modeling Path Dependence in Agricultural Adaptation to Climate Change

Most modeling of agricultural adaptation to climate change is based on some kind of optimization scheme that explicitly or implicitly assumes farmers' clairvoyant knowledge of states of nature, rational behavior, and cost-free decisions. In reality, adaptation by farmers to climate variability and change is impeded by a number of competing factors—including, for example, cost, lack of information, risk tolerance, availability of appropriate adaptation strategies, and resistance to change. Moreover, farmers' initial adaptive choices in response to a dynamic climate can lead to long-term "lock-in" that limits ability to adopt new technologies or strategies to further mitigate negative events, or to take advantage of opportunities. Examples often given of lock-in from other aspects of society include the persistence of the QWERTY keypad, reliance on a single-gauge railroad in the United States, and, more recently, the persistence of analog-signal televisions (David 1985). Lock-in is a form of path dependence, which contends that current outcomes are partly the result of prior decisions. Moreover, decisions made or technological paths pursued during transitory conditions can persist long after those conditions have changed to a new state. Lewandrowski (1993) describes how federal agricultural policy may hinder farmers' ability to adapt to climate change by encouraging the growth of just a few cereal grains within baseline programs that require a farmer to grow a crop for five years before qualifying for support payments. Additional examples of path dependence in agriculture include large capital outlays for items such as machinery and storage facilities, and the market forces that drive agricultural research towards enhancing, rather than stabilizing, yields (Smithers and Blay-Palmer 2001).

Chhetri et al. (2010) assert that failure to account for path dependence in the simulation of agricultural adaptation to climate change produces unrealistic estimates of adaptive success and adds structural uncertainty to model estimates. To address this shortcoming, they simulated the effects of path dependence on corn yields in the southeast United States by imposing alternative assumptions of future change in climate-variance structures and accompanying adaptation on corn production. Chhetri et al. (2010) assert that path dependence of farmers' adaptation decisions during a period of highly dynamic climate increases the risk that those decisions are rapidly rendered ineffective by subsequent climate variability and change. The central question the research seeks to answer is: how is the effectiveness of a suite of adaptation responses to current climate change altered by subsequent changes in mean climate *and* variability when those initial responses are locked in for a long period of time?

Using a modified approach of Easterling et al. (2003), Chhetri et al. (2010) developed a model of adaptation based on the typical epidemiological diffusion function (logistic or "S-curve" of innovation adoption versus time). In this approach, adaptation is not a homogenous temporal process; a minority of farmers will adapt quickly to a perceived change in climate (early adopters), the majority will adapt after two to three years (middle adopters), and a minority will take several years to adapt to the new climate (late adopters). The sum over time of simulated crop yields obtained by deploying these three classes of adopters provides an aggregate measure of adaptive success under this simulated path-dependency restriction. A protocol was followed in which three gradations of path dependence were examined under variable climatic variance: "severe" path dependence (no adaptation), epidemiological path dependence (the logistic model just described), and no path dependence (climate-optimized instantaneous adaptation). A randomized climatic variance structure was applied to the climate scenario through a random distribution of possible future outcomes. The range of possible outcomes extended up to twice the original climatic variability. This protocol allowed for simultaneous evaluation of the effects of path dependence and nonmonotonic changes in future climate on farmers' adaptive capacity.

The study area is the southeastern United States. We used the EPIC crop model (Williams 1984) to evaluate the long-term effects of climate change and adaptation protocol on corn yields. Changes in crop-planting date and cultivar-maturity class were the simulated adaptations. The crop-model runs were performed on each spatial grid cell of the model for every adaptation and climate change scenario. We chose to focus on corn yields, as it is the most widely grown crop in the study area.

The results of the simulations showed that the advantage of instantaneous, climate-optimized adaptation (i.e., no path dependence) declines through time relative to an epidemiological adaptation process (i.e., moderate path dependence). In the first years of climate change, yields in the optimized adaptation scenario are higher than in the moderate path-dependence scenario (while the yields between the epidemiological and no-adaptation scenarios are almost identical). However, by the end of the climate change simulation, the yields between these two adaptation scenarios converge, while the no-adaptation scenario (i.e., severe path dependence) continues to decline precipitously. The importance of uncertain climate variability is underscored by the fact that yields declined equally across all adaptation scenarios and all time segments under the randomized variance structure.

We interpret these results as suggesting that the cost of path dependence is a yield penalty to farmers. The failure to incorporate modeling structures that explicitly account for path dependence contributes structural uncertainty to model estimates.

Structural uncertainty will never be fully constrained in predictive models of agricultural adaptation to climate change. However, steps can be taken by observer-modelers to illuminate the sources of structural uncertainty. This starts by stating the boundary assumptions of models unambiguously, and by communicating the existence of incomplete or competing model structures. For example, most crop models rely on the concept of optimum photosynthetic temperature for the estimation of temperature stress on plants. Yet, significantly different optima for the same crop are used among different models. The difference in optimum photosynthetic temperature between EPIC-maize and CERES-maize models is approximately 1°C, which introduces a built-in bias (figure 6). Such differences make a strong case for the

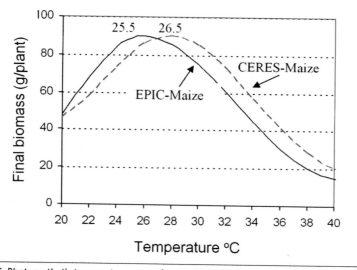

Figure 6. Photosynthetic temperature curves for the EPIC–Maize and CERES–Maize models

conduct of coordinated intercomparison of simulation models with shared model structures and sets of outputs. Eventually, impact- and adaptation-model ensembles should be constructed for use with climate model ensembles.

Value or Parametric Uncertainty

Climate change is in a class of problems for which there is no exact precedent in recent human history. Moreover, the causes and consequences of climate change involve complex systems that range in scale from the local to the planetary. Because of the unprecedented magnitude and extent of the problem, existing environmental and social observation systems have failed to provide the necessary data to fully constrain the interactions between climate variability and change, ecosystems, and society. Even experimental data are inadequate for robust modeling of interactions and feedbacks among the biological and management components of agricultural production models. These deficiencies add *value or parametric uncertainty* to the modeling of agricultural adaptation to climate change. Value or parametric uncertainty stems from inconsistent or incomplete model development and application.

The largest single source of parametric uncertainty in the results of impact and adaptation models derives from the linkage of climate-change-model predictions with impact-adaptation models. The features of climate change scenarios that are most problematic for impact models are the estimated magnitude, rate, variability (in time and space), and scale (temporal and spatial) of climate change (Katz 2002). While all of those features introduce uncertainty for crop modeling, variability and scale are particularly challenging. Crop yields vary much more with changes in interannual climate variability than changes in climate means alone (Gitay et al. 2001). The change in occurrence of extreme events has a much larger impact on Kansas wheat yields, for example, than a simple change in mean climate with no corresponding change in interannual climate variability—i.e., change in the frequency and intensity of extreme events such as droughts and heat waves (Mearns et al. 1996). An example of the effect of manipulating the variance of climate change on the distribution of maize yields across the southeastern United States is shown in figure 7. Were the variance of mean annual temperature to double over its present value as the overall warming proceeds, simulated yield losses suggest major adaptive challenges for farmers across a much larger extent of the southeastern United States than if the variance were to remain unchanged.

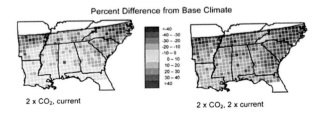

EPIC simulations of maize yield response to different variances of RegCM $2xCO_2$ climate change (1960–95 baseline).

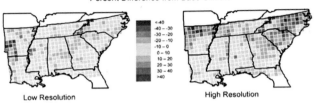

EPIC simulations of maize yield response to low resolution $2xCO_2$ climate change and to high resolution $2xCO_2$ climate change.

Figure 7. Comparison of crop yield simulations forced by low-resolution GCMs with those forced by high-resolution nested regional climate models

The scale resolution of physiological crop models is usually no more than a hectare—that is, the scale of observation of plant growth processes and the interactions of plants with their surrounding environment. However, the scale resolution of GCMs is hundreds to thousands of kilometers. This scale mismatch introduces large uncertainty in simulated crop-yield response to climate change across regions. Spatial-scale uncertainty can be quantified by comparing crop-yield simulations forced by low-resolution GCMs with those forced by high-resolution nested regional climate models (figure 7). In the example shown in figure 7, maize yield response to a coarse-resolution climate model projection of equilibrium climate change is spatially distributed much differently from yield response to a fine-resolution climate model projection.

Scaling/aggregation assumptions introduce another significant form of uncertainty to impact models. Physiological crop models are, for all practical purposes, point-specific—they model growth processes at a particular point in the landscape. Yet their predictions are often treated as representative of regions. Scaling is typically done by weighted averaging of model outputs over areas of homogeneous inputs such as soils or cropping system. Often the model outputs are derived from nonlinear variable-process relationships, which make scaling up tricky. Averaging (making linear) nonlinear relationships over space creates significant aggregation errors illustrated in the

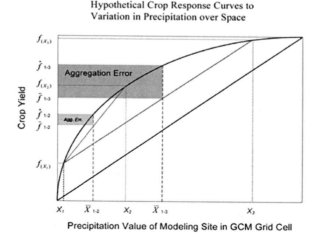

Hypothetical Crop Response Curves to Variation in Precipitation over Space

Figure 8. Aggregation error resulting from averaging values predicted by a nonlinear variable-process relationship—the difference between the linear locus of points described by averaging and that described by the true nonlinear functional relationship is the aggregation error.
Source: Adapted from Easterling and Kok (2002).

abstract in figure 8. The more nonlinear the functional relationship, the greater the aggregation error introduced by averaging.

There are many more sources of parametric uncertainty in impact and adaptation assessments than have been mentioned here, but the above are among the more widely recognized ones. We posit that the representation of parametric uncertainty in adaptation simulations is the most tractable of the three major forms of uncertainty. Rarely have research procedures that use crop simulation models invoked the use of probabilistic techniques such as Monte Carlo simulation in order to constrain the consequences of bias in the input data. The failure to represent modeling results across a range of possible data inputs gives the erroneous appearance that those results are deterministic. Monte Carlo simulation and other bootstrapping techniques are needed to address this problem. Winkler in this volume discusses the use of suites of models to help gauge uncertainty. Such techniques provide useful information for quantifying the uncertainty surrounding the estimates of effectiveness of multiple adaptation strategies.

THE WAY FORWARD: BUILDING A BOND OF TRUST BETWEEN ADAPTATION RESEARCHERS AND FARMERS

It is interesting to speculate about why farmers and other decision makers have paid so much attention to the latest thinking about the science of global warming and so little attention to actionable information generated by the impact-assessment and adaptation-research community. Cascading uncertainty from climate change science to impacts and adaptation surely must be a factor. If that is accepted, then it follows *ceteris paribus* that an initial step toward deliberate coping with the agricultural risks posed by irreversible climate change will require the development of a bond of trust between the adaptation-research community and farmers. That trust must be predicated on clear communication of uncertainties from scientist to stakeholder. Such would seem to be a fundamental requirement in order for a body of information to become usable knowledge. There are many ways to proceed in developing the bond of trust needed to initiate science-based adaptation, but the case study of climate-forecast use by Argentine farmers illustrates one possible approach. Hall and Root (this volume) and Scheraga (this volume) also emphasize dialogue with decision makers. Winkler et al. (this volume) provide an example of a process working with stakeholders, and Moser (this volume) a tool for facilitating such interaction.

How Argentine Soybean Farmers Judge the Utility of Climate Forecasts

With recent advances in seasonal climate forecasting skill, interest has developed in improving the potential for decision making across a range of applications (Bert et al. 2006). Such forecasts are now recognized as potential components of the adaptation toolkit for farmers and other system managers at risk of climate change. Significant challenges exist, however, in the translation of such forecasts into usable knowledge.

The progress that has led to increased forecast skill has not unleashed widespread use of forecasts by decision makers (Podestá et al. 2001; Nicholls 2000). Letson et al. (2001) assert that the assumption that decision makers will use an available forecast explicitly ignores the role of user needs and expectations. Herein is a challenge: understanding users' needs, problems, characteristics, and decision-making contexts to assist the development and provision of relevant and useful climate forecasts. Specifically, climate forecasts must be linked to users' needs and specific contexts before they can be effectively applied to decision making (Hammer 2000a, 2000b; Hammer et al. 2001; Orlove and Tosteson 1999; Stern and Easterling 1999). Recent research indicates that such linkages are best developed when scientists and decision makers work together to develop and design forecast products. User involvement generates credibility and encourages a collaborative approach to design through which climate forecasts are able to reach full potential (Hammer et al. 2001). Furthermore, the relevance and applicability ensured by the creation and strengthening of linkages between climate prediction and user decision making encourages the translation of forecast information into usable knowledge.

Farmers, like most decision makers, routinely make decisions in the face of risk and uncertainty (see discussion by Marx and Weber, this volume). That does not, however, ensure that they know how properly to interpret uncertain information, such as a probabilistic climate forecast (e.g., Hammer 2000a, 2000b; Patt and Gwata 2002). Linkages between climate prediction and user decision making can ease this struggle as well. As scientists, forecasters, and decision makers collaborate to develop forecast products, their interaction can facilitate a multidirectional flow of information that supports processes of reciprocal learning (Hammer 2000a; Podestá et al. 2003; Stern and Easterling 1999). Where interaction serves to inform predictive climate science of users' needs and perspectives, it also encourages users' awareness of prediction capabilities and available information. Interaction provides an arena for clear communication, and the opportunity to develop a bond of reciprocity and trust between science and practice.

Seipt (2006) illustrates the importance of linking soybean farmers and climate forecasters in a formal discourse in the Argentine Pampas. The study investigates how farmers in the Pampas prioritize and trade off different attributes of a seasonal climate forecast (i.e., mechanism of distribution, spatial resolution, lead time, and forecast performance) when judging its utility for on-farm management decisions. Conjoint analysis was used to decompose farmers' holistic evaluations of hypothetical climate forecasts to estimate utility preference structures. Farmers' preference structures were quantified using constrained multiple regression analysis to determine the importance value of each attribute and to determine the utility trade-offs that farmers find acceptable between different attributes. Analysis indicated that the spatial resolution of the forecast, not performance reliability, was the most influential attribute in determining climate forecast utility. Moreover, tradeoff analysis revealed that substantial improvements in other attributes (e.g., lead time) were needed in order to overcome the primacy that farmers assign to spatial resolution in determining the utility of a forecast.

Communication of user preferences to climate scientists and forecasters creates a link between science and agricultural decision making in the region, and encourages the incorporation of such information into future forecast design. The link also serves as a platform for future communication and interaction, thus promoting progress in the translation of the region's seasonal climate forecasts and other climate-change adaptation information into usable knowledge.

CONCLUSION

Most of the scientific information that has been developed to provide insights into the process of adapting agricultural systems to climate change has failed to become usable knowledge to farmers and their supporting institutions. There is a certain paradox in such a statement because much of that information would seem to satisfy the basic requirements of usable knowledge laid out by Haas (2004) and reviewed above, including *tractability, credibility, legitimacy,* and *salience.*

Adaptive information being delivered by researchers is clearly tractable, because many of the recommended adaptive techniques are already in use to manage current climate risk (e.g., diversification of crop species, water conservation, planting of cultivars with specific desired traits). The credibility of adaptive science, notwithstanding well-specified boundary assumptions, is generally quite high in most of the research reviewed in IPCC reports. The IPCC itself adds legitimacy to the body of scientific knowledge on agricultural adaptation. The work is highly salient because it is being delivered by scientists decades ahead of the climate changes anticipated in the climate-model scenarios that serve as test beds for adaptive analysis.

The missing catalyst for action is a vigorous discourse between scientists and farmers over the uncertainties surrounding adaptation research. An interactive dialogue between scientists and farmers promotes communication in both directions. Scientists must be honest brokers of the uncertainties associated with their research findings and recommendations. Farmers and other agricultural stakeholders must assume responsibility for becoming part of the adaptation solution by providing crucial input to the research establishment on what is helpful and what is not. Without this discourse, we are all shooting in the dark.

REFERENCES

Adams, R.M., B.A. McCarl, K. Seerson et al. 1999. The economic effects of climate change on U.S. agriculture. Chap. 2 in *Impacts of Climate Change on the U.S. Economy,* ed. by R. Mendelsohn and J. Neumann. Cambridge: Cambridge University Press.

Bert, F.E., E.H. Satorre, F.R. Toranzo, G.P. Podestá. 2006. Climatic information and decision-making in maize crop production systems of the Argentinean Pampas. *Agricultural Systems* 88:180–204.

Bruinsma, J., ed. 2003. *World Agriculture: Outlook to 2015/2030. An FAO Perspective.* Rome: Earthscan Report.

Chhetri, N. and W.E. Easterling. 2008. Climate change and food security in dryland regions of the world, *Annals of Arid Zone* 47: 1–12.

Chhetri, N., W. Easterling, A. Terando, and L. Mearns. 2010. Modeling path dependence in agricultural adaptation to climate variability and change, *Annals of the Association of American Geographers* 100(4): 894–907.

Chow, J., R. Kopp, and P. Portney. 2003. Energy resources and global development. *Science* 302:1528–1531.

Darwin, R.F., M. Tsigas, J. Lewandrowski, and A. Raneses. *World Agriculture and Climate Change: Economic Adaptations.* Agricultural Economic Report Number 703, U.S. Department of Agriculture, Economic Research Service, Washington, DC, 1995.

David, P. 1985. Clio and the economics on QWERTY. *American Economic Review* 75: 332–337.

Delgado, C., C. Courbois, and M. Rosegrant. 1999. Global food demand and the contribution of livestock as we enter the new millennium. MSSD Discussion Paper 21, International Food Policy Research Institute, Washington, DC. Available at http://www.ifpri.org/divs/mtid/dp/papers/dp21.pdf.

Dyson, T. 1999. World food trends and prospects to 2025. *Proceedings of the National Academy of Sciences* 96(11):5929–5936.

Easterling III, W.E. 1996. Adapting North American agriculture to climate change in review. *Agricultural and Forest Meteorology* 80(1):1–53.

Easterling, W.E. 2006. Climate change, multiple stresses, and agriculture: When all else is not equal. In *Understanding and Responding to Multiple Environmental Stresses*, by Committee on Earth-Atmosphere Interactions, National Research Council, 108–111. Washington, DC: National Academy Press.

Easterling, W.E., P.K. Aggarwal, P. Batima et al. 2007. Food, fibre and forest products. In *Climate Change 2007: Impacts, Adaptation and Vulnerability*, ed. by M. Parry et al., 273–313. Cambridge: Cambridge University Press.

Easterling, W.E., and M. Apps. 2005. Assessing the consequences of climate change for food and forest resources: A view from the IPCC. *Climatic Change* 70:165–189.

Easterling, W.E., N. Chhetri, X. Niu. 2003. Improving the realism of modeling agronomic adaptation to climate change: Simulating technological substitution. *Climatic Change* 60:149–173.

Easterling, W., B. Hurd, and J. Smith. 2004. *Coping with Global Climate Change: The Role of Adaptation in the United States.* Washington, DC: Pew Center on Global Climate Change.

Easterling, W., and K. Kok. 2002. Emergent properties of scale in global environmental modeling: Are there any? *Integrated Assessment* 3(2–3):233–246.

Escobar, J., E. Lora, O. Venturini, E. Yanez, E. Castillo, and O. Almazan. 2009. Biofuels: Environment, technology and food security. *Renewable and Sustainable Energy Reviews* (13):1275–1287.

Fawcett, A.A., K.V. Calvin, F.C. de la Chesnaye, J.M. Reilly, and J.P. Weyant. 2009. Overview of EMF 22 U.S. transition scenarios. *Energy Economics* 31:S198–S211.

FAO. 2002. HIV/AIDS, agriculture and food security in mainland and small countries of Africa. *Proceedings of the Twenty-First Regional Conference for Africa, Cairo.*

FAO. 2005. *World Agriculture Towards 2030/2050: An Interim Report*, Global Perspective Studies Unit, Food and Agriculture Organization of the United Nations, Rome.

Fischer, G., M. Shah, H. van Velthuizen, and F. Nachtergaele. 2001. *Global Agro-Ecological Assessment for Agriculture in the 21st Century.* Laxenburg, Austria: IIASA.

Frank, K.L., T.L. Mader, J.A. Harrington, G.L. Hahn, and M.S. Davis. 2001. Climate change effects on livestock production in the Great Plains. *Proc. 6th Intl. Livest. Envir. Symp., Amer. Soc. Agric. Eng.*, St. Joseph, MI, 351–358.

Gitay, H., S. Brown, W. E. Easterling et al. 2001. Ecosystems and their goods and services. In *Impacts, Adaptation and Vulnerability.* IPCC Working Group II Contribution to the Third Assessment Report on Climate Change. Intergovernmental Panel on Climate Change. Available at http://www.grida.no/publications/other/ipcc_tar/.

Haas, P. 2004. When does power listen to truth? A constructivist approach to the policy process. *Journal of European Public Policy* 11(4):569–592.

Hammer, G. 2000a. A general systems approach to applying seasonal climate forecasts. In *Applications of Seasonal Climate Forecasting in Agricultural and Natural Ecosystems: The Australian Experience*, ed. by G.L. Hammer, N. Nicholls, and C. Mitchell, 51–65. Dordrecht, The Netherlands: Kluwer.

Hammer, G. 2000b. Applying seasonal climate forecasts in agricultural and natural ecosystems: A synthesis. In *Applications of Seasonal Climate Forecasting in Agricultural and Natural Ecosystems: The Australian Experience*, ed. by G.L. Hammer, N. Nicholls, and C. Mitchell, 453–462. Dordrecht, The Netherlands: Kluwer.

Hammer, G.L., J.W. Hansen, J.G. Phillips et al. 2001. Advances in application of climate prediction on agriculture. *Agricultural Systems* 70:515–553.

Hayami, Y., and V.W. Ruttan. 1985. *Agricultural Development: An International Perspective.* Baltimore, MD: Johns Hopkins University Press.

IPCC. 2000. *Special Report: Emissions Scenarios.* A Special Report of Working Group III of the Intergovernmental Panel on Climate Change. Geneva: World Meteorological Organization.

IPCC. 2007. *Climate Change 2007: The Physical Science Basis.* Edited by S. Solomon, D. Qin, M. Manning et al. Cambridge and New York: Cambridge University Press.

ISAAA. URL: http://www.isaaa.org

Karl, T.R., and K. Trenberth. 2003. Modern global climate change. *Science* 302:1719–1723.

Kates, R.W., and W. Clark. 1996. Expecting the unexpected. *Environment* 38(2):6–11, 28–34.

Katz, R.W. 2002. Techniques for estimating uncertainty in climate change scenarios and impact studies. *Climate Research* 20:167–185.

Letson, D., I. Llovet, G. Podestá et al. 2001. User perspectives of climate forecasts: Crop producers in Pergamino, Argentina. *Climate Research* 19:57–67.

Lewandrowski, J. 1993. Farm programs and climate change. *Climatic Change* 23:1–20.

Long, S.P., E.A. Ainsworth, A.D.B. Leakey, J. Nosberger and D.R. Ort. 2006. Food for thought: Lower expected crop yield stimulation with rising CO_2 concentrations. *Science* 312:1918–1921.

Manning, M., M. Petit, D. Easterling et al. 2004. *Describing Scientific Uncertainties in Climate Change to Support Analysis of Risk and of Options.* Intergovernmental Panel on Climate Change Workshop Report. Available at http://www.ipcc.ch/.

Mearns, L.O. 1995. Research issues in determining the effects of changing climate variability on crop yields. In *Climate Change and Agriculture: Analysis of Potential International Impacts.* ASA Special Publication No. 59, Madison, WI, 123–143.

Mearns, L.O., F. Giorgi, L. McDaniel, and C. Shields. 2003. Climate scenarios for the southeastern US based on GCM and regional model simulations. *Climatic Change* 60:7–35.

Mearns, L.O., R.W. Katz, and S.H. Schneider. 1984. Extreme high-temperature events: Changes in their probabilities with changes in mean temperature. *Journal of Climate and Applied Meteorology* 23:1601–1613.

Mearns, L., C. Rosenzweig, and R. Goldberg. 1996. The effect of changes in daily and interannual climatic variability on CERES-wheat: A sensitivity study. *Climatic Change* 32:257–292.

Mendelsohn, R., W. Nordhause, and D. Shaw. 1994. Impact of global warming on agriculture: A Ricardian analysis. *American Economic Review* 84:753–771.

Mendelsohn, R., A. Dinar, A. Basist et al. 2004. Cross-sectional analyses of climate change impacts. World Bank Policy Research Working Paper 3350, Washington, DC.

Moss, R.H., and S.H. Schneider. 2000. Uncertainties in the IPCC TAR: Recommendations to lead authors for more consistent assessment and reporting. In *Guidance Papers on the Cross-cutting Issues of the Third Assessment Report of the IPCC*, ed. by R. Pachauri, T. Taniguchi, and K. Tanaka (annex 2: extended abstracts), 33–51. Geneva: Intergovernmental Panel on Climate Change.

Nicholls, N. 2000. Opportunities to improve the use of seasonal climate forecasts.

In *Applications of Seasonal Climate Forecasting in Agricultural and Natural Eco-systems: The Australian Experience*, ed. by G.L. Hammer, N. Nicholls, and C. Mitchell, 309–327. Dordrecht, The Netherlands: Kluwer.

NRC. 1983. *Changing Climate.* Washington, DC: National Academy Press.

———. 2001. *Climate Change Science: An Analysis of Some Key Questions.* Committee on the Science of Climate Change, pp. 18–21. Washington, DC: National Academy Press.

———. 2002. *Predicting Invasions of Nonindigenous Plants and Plant Pests.* Committee on the Scientific Potential for Predicting the Invasive Potential of Nonindigenous Plants and Plant Pests in the United States. Washington, DC: National Academy Press.

———. 2003. *Communicating Uncertainties in Weather and Climate Information: A Workshop Summary.* Board on Atmospheric Sciences and Climate, pp. 39–42. Washington, DC: National Academy Press.

Orlove, B.S., and J.L. Tosteson. 1999. The application of seasonal to interannual climate forecasts based on El Niño–Southern Oscillation (ENSO) events: Australia, Brazil, Ethiopia, Peru, and Zimbabwe. Institute of International Studies, Berkeley Workshop on Environmental Politics, Working Paper WP99-3-Orlove. Available at http://repositories.cdlib.org.

Parry, M., G. Fischer, M. Livermore, C. Rosenzweig, and A. Iglesias. 1999. Climate change and world food security: A new assessment. *Global Environmental Change* 9(1999 suppl.):S51–S67.

Parry, M.L., C. Rosenzweig, and M. Livermore. 2005. Climate change, global food supply, and risk of hunger. *Philosophical Transactions of the Royal Society B* 360:2125–2138.

Patt, A., and C. Gwata. 2002. Effective seasonal climate forecast applications: Examining constraints for subsistence farmers in Zimbabwe. *Global Environmental Change* 12:185–195.

Pilon-Smits, E.A.H., M.J. Ebskamp, M. Paul, M. Jeuken, P. Weisbeek, and S. Smeekens. 1995. Improved performance of transgenic fructan-accumulating tobacco under drought stress. *Plant Physiology* 107:125–130.

Pingali, P., L. Alinovi, and J. Sutton. 2005. Food security in complex emergencies: Enhancing food system resilience. *Disasters* 29(SI):S5–S24.

Pinstrup-Andersen, P., R. Pandya-Lorsh, and M. Rosegrant. 1999. *World Food Prospects: Critical Issues for the Early Twenty-First Century.* Food Policy Report, International Food Policy Research Institute, Washington, DC.

Podestá, G.P., K. Broad, and D. Letson. 2003. Interannual climate variability and agriculture in Argentina: What did we learn? Background paper for workshop "Insights and Tools for Adaptation: Learning from Climate Variability," organized by NOAA/OGP, Washington, DC, 18–21 November 2003.

Podestá, G.P., D. Letson, J.W. Jones et al. 2001. Aplicación de información climática relacionada con el fenómeno ENOS en el sector agrícola de la Pampa húmeda argentina. VIII Argentine Meteorology Congress and IX Latin American and Iberic Meteorology Congress. Buenos Aires, Argentina, May 7–11, 2001.

Roughgarden, T., and S.H. Schneider. 1999. Climate change policy: Quantifying uncertainties for damages and optimal carbon taxes. *Energy Policy* 27(7):415–429.

Ruttan, V.W. 1997. Induced innovation, evolutionary theory and path dependence: Sources of technical change. *Economic Journal* 107:1520–1529.

SAP. 2008. Climate change: Challenges and opportunities for business. SAP white paper. Available at http://www.sap.com/community/showdetail.epx?ItemID =15250.

Seipt, E.C.K., 2007. Understanding Argentine farmers' perceptions of the utility of seasonal climate forecasts. Master's thesis, Department of Geography, Pennsylvania State University.

Slingo, J., A. Challinor, B. Hoskins, and T. Wheeler. 2005. Introduction: Food crops in a changing climate. *Philosophical Transactions of the Royal Society B* 360:1983–1989.

Smit, B., I. Burton, R.J.T. Klein, and J. Wandel. 2000. An anatomy of adaptation to climate change and variability. *Climatic Change* 45:223–251.

Smithers, J., and A. Blay-Palmer. 2001. Technology innovation as a strategy for climate adaptation in agriculture. *Applied Geography* 21:175–197.

Stern, P.C., and W.E. Easterling. 1999. *Making Climate Forecasts Matter.* Washington, DC: National Academy Press.

Tilman, D., J. Fargione, B. Wolff et al. 2001. Forecasting agriculturally driven global environmental change. *Science* 292:281–284.

Tubiello, F.N., J.A. Amthor, K. Boote, M. Donatelli, W.E. Easterling, G. Fisher, R. Gifford, M. Howden, J. Reilly and C. Rosenzweig, 2007. Crop response to elevated CO_2 and world food supply. *European Journal of Agronomy* 26:215–223.

Williams, J.R., C.A. Jones, and P.T. Dyke. 1984. A modeling approach to determining the relationship between erosion and soil productivity. *Transactions of the ASAE* 27:129–144.

Ziska, L.H. 2003. Evaluation of yield loss in field-grown sorghum from a c3 and c4 weed as a function of increasing atmospheric carbon dioxide. *Weed Science* 51:914–918.

Adapting to Climate Change in the Context of Multiple Risks

A Case Study of Cash Crop Farming in Ontario

BEN BRADSHAW, SUZANNE BELLIVEAU,
AND BARRY SMIT

AGRICULTURE IS INHERENTLY SENSITIVE TO CLIMATIC CONDITIONS, and hence is frequently cited as a sector that is potentially vulnerable to anticipated global climate change; indeed, numerous scenario-based climate-change impact assessments have forecast problematic effects such as declining crop yields and heightened food insecurity (e.g., Rosenzweig 1990; Rosenzweig and Parry 1994; Brklacich and Stewart 1995). These concerns are reinforced by other works in this volume (Andresen; Easterling et al.; Winkler et al.) that shed light on the potential implications of climate change for agriculture in the Great Lakes region, specifically. Of course, the degree to which an agricultural system is ultimately vulnerable to long-term changes in temperature or precipitation, or changes in the frequency and magnitude of extreme weather events, depends on its adaptive capacity; it is for this reason that the literature on climate change and agriculture has increasingly directed attention to the issue of adaptation. (For reviews, see Tol et al. 1998; Smit and Pilifosova 2001.) While it is widely recognized that farmers have the ability to reduce the adverse effects of climate change and even seize opportunities by adapting to the changing conditions (Easterling 1996; Wheaton and MacIver 1999; Bryant et al 2000; Smit and Skinner 2002), the exact process through which this adaptation will occur, or indeed even does occur at present, is not well understood (Brklacich et al. 1997; Bryant et al. 2000; Lemmen and Warren 2004).

One significant complication is the impact of non-climatic stimuli on adaptation decisions, as climate obviously represents just one of many sources of risk to which farmers are exposed and respond. Events such as commodity market downturns, changes to government support programs, fluctuations in currency and interest rates, highly contagious livestock diseases, or the loss of export markets due to consumer health concerns may present significant risks to producers at certain times. Even upswings in prices, as has recently been experienced by Great Lakes cash-crop producers given increasing demand for biofuels and livestock feed, can create new risks, given the tendency of producers to specialize and overcapitalize (Hurt 1981; Bradshaw et al. 2004). More problematically for researchers, it is highly likely that these multiple stimuli interact to influence farmers' decisions, including their farm management practices, and hence agricultural adaptations to climatic variability and change cannot be conceived via simple single stress–single response models (Risbey et al. 1999; Smit et al. 2000; Bradshaw et al. 2004; Adger et al. 2003).

Another barrier to improving our understanding of the likely adaptive response of farmers to anticipated climate change derives from the possibility that farmers will experience and respond to what Yohe (2000) labels "inter-periodic variability" rather than long-term climate change. While a healthy debate persists as to whether or not this is problematic for farmers' future adaptive capacity (e.g., see Schneider et al. 2000; Mendelsohn 2000 for the "yes it is" perspective and Downing et al. 1997; Yohe 2000 for the "no it is not" perspective), more significantly this speaks to the difficulty of predicting the likely farm-level response to anticipated climate change.

A third and more fundamental barrier to improved knowledge of climate change adaptation derives from the simple fact that humans can respond in highly variable ways to similar external stimuli (Bryant et al. 2000; Kandlikar and Risby 2000; Marx and Weber, this volume). This reality of human nature likely represents the greatest test for scenario-based climate-change impact assessments that attempt to incorporate adaptation based on heroic assumptions of human behavior (e.g., Easterling et al. 1993; Easterling et al. this volume; Rosenzweig and Parry 1994), leading Schneider et al. (2000) to suggest that future impact assessments must identify—that is, make explicit—their assumptions regarding human adaptive behavior, and even offer multiple outputs based on various assumptions.

In light of these and other barriers to improving our understanding of the likely adaptive response of farmers to anticipated climate change, many in the field have called for empirical assessments of actual adaptive behavior in particular places over particular periods of time (e.g., Smit et al. 1996; Smithers and Smit 1997; Bryant et al. 2000; Kandlikar and Risbey 2000; Polsky and Easterling 2001), even though such behavior is place- and time-specific, and likely represents a response to interperiodic climatic variability, as well as to multiple non-climatic risks and opportunities. In an agricultural context, such case studies, or what Tol et al. (1998) label "temporal analogues," are relatively few. For example, Smithers and Smit (1997) drew on aggregate data representing thousands of producers to look for covariations in climate stimuli, cropping areas, crop yields, crop insurance, and technology over three decades in southern Ontario, and determined that climate stimuli accounted for limited change in cropping areas, due in large part to the presence of crop insurance. In the cases of Smit et al. (1996), Smit et al. (1997), and Brklacich et al. (1997), a limited number of farmers in different regions in Ontario were surveyed to document specific changes in farm practices over a previous period and the reasons offered for such changes. In all cases, results revealed that some farmers had undertaken tactical and strategic changes in light of climatic stimuli, especially annual conditions, but these changes also reflected the risks and opportunities presented by economic, technological, social, and political factors.

The "temporal analogue" or case study of climate change adaptation reported in this chapter, which focuses on cash-crop farmers in Ontario's Grand River Watershed, follows the more intensive approach of Smit et al. (1996), Smit et al. (1997), and Brklacich et al. (1997)—that is, it draws upon the recent experiences of a limited number of producers, documenting considerable information for each. Its overall aim is to better understand producers' adaptation to climatic variability and change in the context of the multiple-risks environment in which they operate. More specifically, the objectives of this chapter are twofold: (1) to identify the climatic risks currently deemed problematic by cash-crop farmers in Ontario, and their efforts to respond to them; and (2) to examine cash-crop farmers' experience with, and response to, other non-climatic risks, and the ways in which this serves to constrain or enhance their capacity to adapt to climate variability and change.

Four additional points are then examined in this chapter. In the next section, further details of the approach to assessing farm-level adaptation to climatic variability and change are offered, along with some background to the case-study location. Section 3 discusses the climatic risks that producers identified as problematic, and some of the adaptive strategies employed to manage these risks. Section 4 examines the adaptations that producers have implemented in response to broader, non-climatic forces, and how these influence producers' capacity to cope with or adapt to climatic risks, particularly those identified in the previous section. Finally, some conclusions are offered.

STUDY SITE DESCRIPTION AND RESEARCH APPROACH

The Grand River Watershed is located in southwestern Ontario, Canada, lying between Georgian Bay and Lake Erie (figure 1). With an overall length of 190 kilometers and a 6,965 square-kilometer drainage area, it is one of the largest watersheds in southern Ontario (GRCA 2006). Agriculture is the dominant land use, with 67 percent of the total land being actively farmed on approximately 6,400 farms (Statscan 2001). The Grand River basin includes parts of 11 counties, with Brant, Waterloo, and Wellington counties being almost fully enclosed; it is within these three counties that cash-crop producers, those growing a combination of soybean, corn, winter wheat, and/or hay, were solicited to understand better their adaptation to climate change in the context of multiple risks.

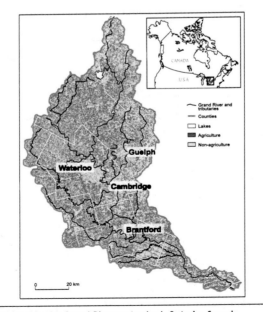

Figure 1. Agricultural land in the Grand River watershed, Ontario, Canada

The research approach used to assess producers' adaptation was guided by the so-called "vulnerability framework" outlined by Ford and Smit (2004) and Smit and Wandel (2006). The approach aims first to identify any and all risks deemed relevant by producers themselves, including but not limited to climatic ones, and then to query producers as to their adaptation to said risks; the combination of their risk exposure and their adaptive capacity is thought to constitute their overall vulnerability. Similar to Belliveau et al. (2006b), the researchers used a combination of interviews and focus-group meetings as primary data sources. Semi-structured interviews were held with 20 cash-crop farmers, who were identified through snowball sampling. The majority of interviewees (75 percent) were second- or third-generation farmers, and the remaining quarter had been farming for a minimum of 11 years.Again, a key aspect of the vulnerability approach is that it does not prioritize any one source of vulnerability, such as weather, but rather tries to identify the risks or exposures that agents themselves deem significant. This need poses some challenges for researchers trying to secure and then query subjects. One approach that has been found to work is to ask producers to recount or review, year-by-year for somewhere between 5 and 15 years, all the challenges they experienced and the strategies they adopted to manage those challenges (Belliveau et al. 2006b). For this case study, 15 years was deemed viable given the age and permanency of the interviewees. In addition to querying producers about yearly challenges and their associated responses, producers were also asked if any broad changes had been made to their operations in that time period (e.g., purchasing or renting new land), what had prompted these changes, and if they had followed an overall risk-management strategy; this was done because many strategic adaptations that occur over more than one year might not have been identified through the previous information solicitation. The final section of the questionnaire focused on just climatic conditions, inquiring about producers' views of climate change and their perceived ability to adapt if problematic conditions became more frequent. In addition to these individual interviews, two focus groups were subsequently held with nine producers who were recruited using signage at a cash-crop producers' trade show in Guelph, Ontario, in 2006. All participants grew a majority combination of corn, soybeans, and winter wheat. Focus-group participants were provided with a summary of the risks and adaptations identified in the interviews as well as historic trends and future projections of climatic conditions. Participants were then asked to discuss the effectiveness of current risk-management strategies, and their likely effectiveness to deal with future climatic conditions, especially in light of the multiple risks that producers experience.

CLIMATIC RISKS AND RESPONSES IN
ONTARIO'S CASH CROP SECTOR

Cash-crop farmers in the Grand River Watershed identified several climate-related conditions that were problematic for their operation, one of which was drought, which occurred in 1988, 1998, 2001, and 2003. Farmers stressed that it was not only the intensity of the drought but also its timing that was important. For corn, yields are most significantly affected if the plant experiences moisture stress during tasseling; this generally occurs around 80 days after planting, which for the study area is around July 10 to 20. In the 2005 growing season, most interviewees reported that the corn crop was near death, due to the hot and dry conditions, until it rained on July 15th, which resulted in record-high corn yields. This example speaks to the difficulty of predicting the likely impacts of anticipated climate change on agriculture, given that the within-season timing of events like precipitation, which cannot be anticipated in advance, largely determines the significance of the climatic risk.Even if a sustained drought occurs, producers have some tools at their disposal with which to respond. Indeed, most interviewees noted that their primary response to drought events is to make a crop insurance claim. Eighty-five percent of respondents carried crop insurance on at least one of their crops, although at least 15 percent of them had only begun purchasing insurance within the last six years in response to changes in the program or in response to a poor crop year. The insurance gives them a buffer to cover the cost of inputs used in a year; however, with each claim, the premiums increase the following year, and the average yield upon which payouts are made decreases. Hence crop insurance is less effective for recurring droughts (or other perils), which may become more frequent under climate change.

The cash-crop producers also identified certain proactive or anticipatory management practices as effective in managing drought risks; that is, many producers have adopted practices that help to maintain production if a drought event occurs. One such practice is conservation tillage, which a majority of respondents use to conserve moisture in the soil, as well as to reduce input costs, minimize soil erosion, improve soil conditions, and enable earlier and quicker planting of crops in the spring. Conservation tillage has the added benefit of reducing the emissions of greenhouse gases that contribute to climate change (Robertson et al. 2000). Zero tillage, which is an extreme form of conservation tillage, is not as widely used with corn crops, in part due to the greater amount of trash produced by the plants that then needs to be worked into the soil. Producers also reduce drought risks through the choice of crop variety. Many respondents tended to choose more drought-tolerant varieties, and varieties that perform well on certain soil types. They also indicated that they are better able to handle drought now than 20 years ago because of the greater variety of seeds.

A greater problem associated with the drought years was the outbreaks of aphids in 2001 and 2003. The combination of drought and aphids cut soybean yields by 40 percent or more in some cases. In addition to reducing yields, the insects reduced the quality of the crop, either preventing them from maturing properly or blemishing the beans; as the beans are graded visually, blemishing results in discounted prices for an output that, even in good years, seldom garners good returns. Aphids were particularly problematic because, in 2001, they were a new pest in the area; in other words, farmers had no experience with them—indeed, they even failed to properly identify them—and were therefore poorly equipped to adapt. Their only response was to claim crop insurance. This finding is consistent with Marx and Weber's (this volume) discussion of how humans cope with uncertainty, and the tendency of decision makers to rely on experience, rather than description, when making decisions. In 2003, farmers were still generally unprepared because of the limited information available to them about threshold levels of pests, preferred spraying practices, and the economic viability of the practices. Since these two incidents, farmers have increased their awareness of the pest and preparedness; however, as there has yet to be a recurrence, producers' enhanced capacity to manage an aphid outbreak has yet to be tested. With the expected rise in winter temperatures (Andresen, this volume), new pests may be able to overwinter, creating potential challenges for Canadian farmers (Lemmen and Warren 2004; see also Hall and Root, this volume, regarding ecological effects of climate change). Unfortunately, the aphid-outbreak example suggests that any climate-related conditions with which farmers have little previous experience will be particularly troublesome, at least initially.

On the other end of the moisture extremes, excessive rain was repeatedly identified by cash-crop producers as problematic, particularly in the spring; this was true of the years 1996 and 2000–2004. Spring conditions are important because there is a narrow window in which planting can occur; if this window is missed, farmers are typically left with unseeded acreage. Depending on the location, farmers ideally plant their corn seeds by May 15th and soybeans by the end of May, or by June 10th at the very latest. If conditions are too wet, then the heavy planting equipment tears up the field, compacts the root zone, and/or becomes stuck in the muddy fields. If seeds are already planted, they may be subjected to soil-crusting problems (which inhibit the plants from pushing through the ground), flood damage to seedlings, or the loss of nitrogen as a result of leaching.

The predominant response of producers to excess rain in the springs of 1996 and 2000–2004 was to delay planting until the fields were dry, which takes longer where soils are poorly drained. However, the longer one waits to plant, the fewer the heat units that are potentially available in a season, which can be expected to reduce yields, especially with corn. This is clearly an issue for southern Ontario corn growers given hypercompetition from producers in the U.S. corn belt, where growing seasons are longer. Delayed planting, of course, also delays the achievement of certain growth stages of plants, such as the silking of corn, which indicates that the crop is near maturity. If silking occurs too late in the growing season when cool nighttime temperatures return, photosynthetic activity and grain filling can be disrupted. Delayed planting also typically results in delayed harvesting. Although there is a bigger window at harvest than at planting, if the harvesting of soybeans is delayed, then so too is the planting of winter wheat, which is often the next crop in farmers' rotations; the later that wheat is planted, the higher the risk of getting a poor crop.

So problematic are wet springs that a majority of the interviewees stated a preference for drought conditions in the spring over very wet conditions, not only because drought conditions allow for earlier planting and a longer season, but also because they prompt deeper root growth, thereby preparing the plant for potential droughts later in the season. Evidently, cash-crop producers are sensitive to intense rainfall in the spring. Problematically for them, a warmer and wetter climate is projected for some areas of the Great Lakes Region (Andresen, this volume).

In addition to delaying planting, another common response to wet springs is to change seed varieties. In some cases, farmers will change to a shorter-season, lower heat-unit corn, which typically yields less, but nonetheless produces a crop. Exchanging seeds with the seed dealer comes at no cost to the farmer, but is only possible when and where alternative seed is available. In other cases, especially when the planting window for corn narrows, farmers will switch from corn to soybean because it has a longer planting window and is less likely to have yield losses. A downside of this practice is that it means that soybeans may be planted on the same field two years in a row, which breaks the rotation schedule and increases the plant's sensitivity to diseases and pests. Further, the no-till practices that assist in drought years can be disadvantageous in wet springs. By leaving the previous crops' trash on the field, it locks in moisture and covers the soil so that it does not warm up as quickly, which further prolongs seeding. However, several producers noted that these "backwards" springs typically occurred just once every five years or more, and so they would not change their practices because of it. The expectation that these probabilities will be altered by climate change has yet to be factored into producers' risk-management strategies.

In addition to wet springs, cold and wet seasons were generally deemed problematic by producers; as one producer put it: "A hot, dry summer will scare you, but a wet, cool summer will starve you." Such conditions, which were observed by interviewees in 1992, 1996, 2000, and 2004, fail to provide the necessary heat units for a crop to grow and mature, or result in growth abnormalities. For example, cool and wet conditions can cause soybeans to flower improperly, so that the plant grows without pods; these same conditions increase the risk of fusarium outbreaks in wheat, particularly if it is wet during flowering at the end of May. Wet falls are also problematic in that they raise the moisture content in crops, which increases drying costs at the elevators and can even cause grain to sprout on the stalk. All of these factors reduce the quality of the crop, lowering its grade and the price received for it. Most significantly, there are few if any practices that farmers can employ to enhance quality. If crops are reduced to the lowest grade, farmers must sell them as feed, for which the lowest possible prices are garnered; in some instances, such as when wheat is contaminated with fusarium, producers may not even be able to find a feed mill willing to take the crop.

Finally, cash-crop producers in the Grand River Watershed identified extended periods of high temperatures as problematic given their effect on crop development. For example, excessive heat can result in sun scald, flower abortion, and reduced pod set in soybeans, problems for which few adaptations are available. For those cash-crop producers with livestock, excessive heat can also cause the animals to suffer from heat stress. For these producers, an adaptation is available, but it is a costly one; two such farmers indicated that after losing livestock to heat stress, they installed cooling systems in their barns. This may be an adaptation that will become more common as temperatures increase and heat waves become more frequent.

The climate-related risks identified above were those that the interviewed producers themselves deemed most relevant. Whether excessive rain in the spring or prolonged drought during the summer, all of these problematic climatic conditions, many of which are expected to increase in frequency and/or intensity under climate change (Andresen, this volume), were implicitly recognized as problematic because of their eventual impact on farm profitability. Reduced yields and/or quality, or added drying costs result in decreased net revenues, and so it is no wonder that producers actively seek to minimize these risks. In some instances, producers are already employing effective adaptation strategies; in other instances, however, few if any effective adaptations are available. A further concern, which emerges from the recognition that cash-crop farmers operate in a multiple-risk environment, stems from the possibility that certain known adaptations are constrained by non-climatic risks and pressures. Or put another way, producers' necessary response to powerful non-climatic risks may affect their ability to adapt to climatic risks now and in the future.

NON-CLIMATIC RISKS AND THEIR IMPACT ON CLIMATE CHANGE ADAPTATION

Although climatic risks are clearly important to cash-crop producers, they are by no means the only risks with which they must deal. The non-climatic risks deemed most problematic by producers were economic ones, with particular concern expressed about the variable but generally low prices received for their crops at the time of the interviews, and the growing costs to grow, store, and sell them. While this "price-cost squeeze" is an age-old phenomenon in commercial agriculture (e.g., see Rochester 1940), the "squeeze" had accelerated for these producers in 2006 due to an appreciating Canadian dollar, which had made sales to export markets more difficult, and spiraling costs for fuel and equipment. On the price side, the past two decades have indeed been a difficult period for Ontario's cash-crop farmers. Using corn as an example, figure 2 identifies the average annual per bushel price, in both nominal and real dollar terms, for the period 1915 to 2005. While nominal prices show some improvement over time, the real price of corn, accounting for inflation, has been highly volatile and highly depressed; in terms of 1992 dollars (Canadian), the price of corn in 2005 was the lowest for the last century.

In response to this pressure, especially in years of true hardship like 2005, producers cited certain short-term or tactical adaptations, such as holding off on purchasing equipment, storing grain in elevators until prices are high, cutting back on inputs like fertilizers, and carefully monitoring commodity prices in order to lock in a portion of their crop when prices are modest to high. Another common response to financial hardship, which was previously identified in the context of conserving soil moisture, is the adoption of no-till or at least conservation tillage. By running tractors less regularly across their fields, producers save both fuel and labor costs,

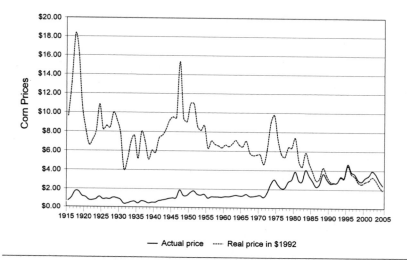

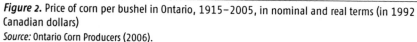

Figure 2. Price of corn per bushel in Ontario, 1915–2005, in nominal and real terms (in 1992 Canadian dollars)
Source: Ontario Corn Producers (2006).

although their use of tillage-replacing herbicides such as glyphosate can increase. While riding out financial difficulties appears feasible and even potentially beneficial with respect to the soil-moisture conservation benefits of conservation tillage, years marked by poor prices and negative net returns can also limit the capacity of producers to manage certain risks associated with climatic variability and change. For example, interviewees universally identified irrigation as an ideal strategy to manage drought conditions, but also universally acknowledged its infeasibility with corn selling for less than 4 dollars (Canadian) per bushel. While some irrigation was evident where farmers grew higher-valued fruits, such as strawberries, irrigation systems were never found on cash-crop fields because the value of the crops failed to justify the investment.

Investing in irrigation constitutes what the climate-change adaptation literature refers to as a long-term management or strategic adaptation. While irrigation may not be feasible for Ontario's cash-crop industry, producers' longtime exposure to the price-cost squeeze has elicited a number of other strategic adaptations. The majority of these were undertaken for economic rather than climatic reasons, but they in turn have implications for producers' capacities to manage climatic risks. One common strategic response was to diversify by, for example, augmenting the number of crop types, taking on livestock, selling alternative agricultural products and services, or securing income from off the farm. With respect to augmenting the number of crop types, only a few farmers had done so beyond the traditional complement of corn, soybeans, and either wheat or hay, choosing to produce such things as strawberries, other fruits, and Identity-Preserved soybeans, which can be segregated by variety, quality, GMO/non-GMO, or specialty trait (e.g., high protein, high sugar, etc.) and therefore benefit from premium prices. To a large degree, cash-crop farmers have already achieved sufficient diversification to reduce market and climatic risks. In this regard, for example, rotating with corn, soybeans, and either wheat or hay breaks disease cycles and ensures that at least one crop will achieve good growth regardless of climatic conditions; winter wheat grows well in cooler seasons and can better tolerate damp conditions, whereas corn and soybeans require a lot of heat and sunlight but need moisture at different times in the growing season.

A more aggressive form of diversification is to add livestock to the operation, including chickens, dairy, beef, hogs and/or sheep; this strategy was adopted by almost half of the interviewees. While costly to get into, chicken and dairy production in Ontario offers considerable security of revenue for producers owing to legislatively backed limits on growth in supply, coupled with onerous tariffs on foreign imports. As a result, these producers have considerably greater capacity to cope with reduced yields and/or prices for their cash crops. Holding livestock can also moderate income variability from year to year because in years when crop quality is compromised due to climate-related stresses and/or prices are low, more of the crop can be used as feed, thereby reducing input costs. Conversely, in years when crop quality is good and/or prices are high, more of the crop can be sold, and feed can be purchased elsewhere or stored feed can be used. Notwithstanding the apparent benefits of such mixed operations, a small proportion of the interviewees had chosen to drop livestock in favor of just cash-crop production. In other words, this form of diversification was not universally viewed as beneficial, or perhaps it was not financially viable for some producers.

The most common form of diversification adopted by the cash-crop producers, which producers explicitly recognized as an effective risk-management strategy, was securing other agricultural or nonagricultural sources of income. Indeed, this strategy was employed by 90 percent of interviewees, representing operations of all sizes. Even the farmer from the largest operation (greater than 5,000 acres) noted that "The farm on its own cannot sustain itself; it's not viable." Additional income was secured through, for example, operating grain elevators (storage and drying facilities) and seed or chemical dealerships; running a bed and breakfast; undertaking custom work such as planting, harvesting, snowplowing, construction, and land development; and securing an off-farm job. Operating a grain elevator, in particular, was viewed as beneficial in that it allows farmers, primarily of larger operations, to capitalize on poor-weather years by drying other producers' crops and reducing their own costs for drying. Additionally, these operators are able to store producers' crops when prices are low for future sale when prices rebound, or store producers' poor-quality crops for later blending with higher-quality crops.

While the interviewees identified many benefits of diversification, many also identified certain drawbacks to this adaptation strategy. For example, with a short window in the spring for planting seeds, working another job or custom-seeding another farm may cause a producer to plant in less than ideal conditions. Even more problematically, when diversification acts to moderate income variability over time, as it is intended to do, producers who partake in the nationally organized Canadian Income Stabilization Program typically find it harder to secure payouts in years marked by low commodity prices or limited farm-derived net income; the program only makes payouts to producers when their whole farm income, including that derived from alternative agricultural services or off-farm jobs, drops significantly below a five-year running average. In other words, current government policy is likely limiting the adoption of a strategy that is known to assist farmers in managing risks associated with climatic variability and change.

Another important long-term management or strategic adaptation to the problem of declining prices and increasing costs undertaken by the interviewees has been to increase the scale of their operations, which in a cash-crop context means securing additional farmland—most typically via leasing, given the high cost of land. Among the 20 cash-crop producers interviewed, half had either secured, or were seeking opportunities to secure, additional land. In addition to achieving cost savings through economies of scale by gaining access to additional fields, especially those at some distance from home fields, the risk of suffering entire crop losses is reduced since weather and soil conditions vary over the region. However, this form of risk-spreading can also limit the utility of crop insurance; because producers purchase coverage for the entirety of each crop type across any and all fields, poor yields from any single field are aggregated with the yields of other fields in order to determine if a claim is justified. A further limitation to the expansion strategy where additional lands are secured through leasing stems from the unwillingness of producers to invest in land improvement such as tile drainage, which represents an effective tool for reducing ponding in wet years and enabling earlier planting generally.

Producers suggested that investments in tile drainage were only viable on owned land, or for lands under lease for no less than 10 years. Where short-term leasing has occurred, which is true in some part among all 20 producers interviewed, a further disincentive to adopt suitable adaptations to wet conditions is evident. In a related vein, short-term leases were found to discourage the adoption of practices aimed at maintaining long-term soil health. While soil health was said to be an important factor in increasing the resilience and robustness of plants to climate-related stresses, where land was managed under short-term leases, interest in short-term gains typically led to inappropriate soil practices. Hence, as with the example of government subsidy programs that limit the financial incentive to manage income variability via output diversification, non-climatic stimuli can influence the vulnerability of producers to current climatic conditions, and certainly can be expected to influence their vulnerability to climate change in the future.

SUMMARY AND CONCLUSIONS

In light of certain challenges to understanding better the likelihood and nature of adaptation to future climate change within agriculture, and especially the problem posed by the multi-risk environment in which farm-level decisions are made, contributors to the scholarship on climate change adaptation have increasingly called for empirical assessments of actual adaptive behavior in particular places over particular periods of time (e.g., Smit et al. 1996; Smithers and Smit 1997; Bryant et al. 2000; Kandlikar and Risbey 2000; Polsky and Easterling 2001). This chapter has reported on one such assessment or "temporal analogue," to use the terminology of Tol et al. (1998). Drawing on the recent experience of cash-crop farmers in Ontario's Grand River Watershed, the case study served to identify the climatic risks currently deemed problematic by the producers, and their efforts to respond to them, as well as to examine producers' exposure and response to other non-climatic risks, and the ways in which these serve to constrain or enhance their capacity to adapt to climate variability and change.

The study found that the climatic stresses that are most problematic are moisture extremes (both too little and too much), cool growing seasons, fusarium and other diseases, new pests, and extreme heat events. Most significantly, the timing of these stresses throughout the growing season determines the severity of the effect. For example, the timing of precipitation largely determines corn yields, with the greatest impact occurring in

mid-July if a period of drought persists. In the case of excessive precipitation in the spring, farmers are forced to delay planting, which shortens the
season and may reduce yield. Beyond yield effects, excess moisture can also
lead to reduced crop quality, which typically results in lower prices for the
crop at harvest time. In response to these and other climatic risks, producers employ a range of management practices. In general, when the risk is
well known, as is the case with droughts, producers are highly adaptive, as
is manifested in the drought example by the use of conservation tillage and
crop insurance. In contrast, when the risk is less familiar, as was the case
with an aphid outbreak, producers are less able to manage the problem.
Of course, climate change is expected to deliver more frequent and severe
droughts and outbreaks of novel pests, which will test producers' adaptive
capacity, particularly with the reduced crop-insurance payouts following
repeated claims.

The cash-crop producers were also found to be actively experiencing
and responding to a number of non-climatic risks, which clearly influenced
their ability to adapt to climatic variability and change. The persistence
of the "price-cost squeeze," for example, is not only pressuring cash-crop
farmers to continually evolve their operations through diversification and/
or farm enlargement, but is also directly constraining the number of adaptation options available to them to manage climatic risks; highlighting the
clearest example of this, none of the producers interviewed for this study
had even considered irrigating their cash crops given insufficient prices and
profit margins. For those strategic responses that have been adopted, their
implications for managing climatic risks now and in the future were found
to be mixed. In the case of diversification via augmenting the number of
crop types, for example, adaptive capacity appears to be enhanced, as different crops tend to be susceptible to different climatic conditions. Similarly,
diversifying income sources provides greater stability in the face of climatic
and/or market risks; however, this form of diversification can also take away
from producers' ability to manage their own farms and may negate opportunities to claim state farm support, given the use of subsidy determination
formulas that include all sources of income. In the case of farmland expansion to achieve greater economies of scale, producers are also self-insuring by
spreading risk geographically; however, this strategy can be counterproduc-
tive when claiming crop insurance given the pooling of crop yields across
all production sites. Lastly, the common use of short-term leases to enable
this expansion tends to inhibit the adoption of well-known adaptations to
climatic extremes such as tile drainage.

These last few examples only begin to touch upon the complications inherent in farm-level decision making in a multi-risk environment, and the related challenge to analysts seeking to understand human adaptive behavior for the purpose of identifying the likely implications of climate change in agriculture. In the interest of making such projections increasingly robust, continued attention to current farm-level adaptive behavior in light of climatic and non-climatic risks appears to be well justified. More productively, efforts are clearly needed to augment adaptive capacity in the cash-crop sector via industry and government policy initiatives. Rather than develop new "climate change adaptation" programs, their efforts should be incorporated into, or at least coordinated with, existing support programs, as there is little evidence that suggests that producers separate their adaptation to climatic risks from that undertaken to manage other risks of production and marketing in Canadian agriculture.

REFERENCES

Adger, W.N., S. Huq, K. Brown, D. Conway, and M. Hulme. 2003. Adaptation to climate change in the developing world. *Progress in Development Studies* 3(3):179–195.

Belliveau, S., B. Smit, and B. Bradshaw. 2006a. Multiple exposures and dynamic vulnerability: Evidence from the grape and wine industry in the Okanagan Valley, British Columbia, Canada. *Global Environmental Change* 16:364–378.

Belliveau, S., B. Bradshaw, B. Smit, S. Reid et al. 2006b. *Farm-Level Adaptation to Multiple Risks: Climate Change and Other Concerns.* Department of Geography Occasional Paper #27, University of Guelph (ISBN 0–88955–558–3).

Bradshaw, B., H. Dolan, and B. Smit. 2004. Farm-level adaptation to climatic variability and change: Crop diversification in the Canadian prairies. *Climatic Change* 67(1):119–141.

Brklacich, M., and R. Stewart. 1995. Impacts of climate change on wheat yields in the Canadian Prairies. In *Climate Change and Agriculture: Analysis of Potential International Impacts*, ed. by C. Rosenzweig, L. Allen, S. Hollinger, and J. Jones, 147–162. Special Publication No. 59. Madison, WI: American Society of Agronomy.

Brklacich, M., D. McNabb, C. Bryant, and J. Dumanski. 1997. Adaptability of agriculture systems to global climate change: A Renfrew County, Ontario, pilot study. In *Agricultural Restructuring and Sustainability*, ed. by B. Ilbery, Q. Chiotti, and T. Rickard, 185–200. London: CAB International.

Bryant C.R., B. Smit, M. Brklacich et al. 2000. Adaptation in Canadian agriculture to climatic variability and change. *Climatic Change* 45:181–201.

Downing, T.E., L. Ringius, M. Hulme et al.1997. Adapting to climate change in Africa. *Mitigation and Adaptation Strategies for Global Change* 2: 19–44.

Easterling, W.E. 1996. Adapting North American agriculture to climate change in review. *Agricultural and Forest Meteorology* 80: 1–54.

Easterling, W., P. Crosson, N. Rosenberg, M. McKenney, L. Katz, and K. Lemon. 1993. Agricultural impacts of and responses to climate change in the Missouri-Iowa-Nebraska-Kansas (MINK) region. *Climatic Change* 24:23–61.

Ford, J., and B. Smit. 2004. A framework for assessing the vulnerability of communities in the Canadian Arctic to risks associated with climate change. *Arctic* 57:389–400.

Grand River Conservation Authority (GRCA). 2009. http://www.grandriver.ca/.

Hurt, R. 1981. *The Dust Bowl: An Agricultural and Social History*. Chicago: Nelson-Hall.

Kandlikar, M., and J. Risbey. 2000. Agricultural impacts of climate change: If adaptation is the answer, what is the question? *Climatic Change* 47:325–352

Lemmen, D.S., and F.J. Warren, eds. 2004. *Climate Change Impacts and Adaptation: A Canadian Perspective*. Ottawa: Natural Resources Canada.

Mendelsohn, R. 2000. Efficient adaptation to climate change. *Climatic Change* 45:583–600.

Ontario Corn Producers Association (2006). Average Ontario grain corn prices. Available at http://www.ontariocorn.org/facts/prices.html.

Polsky, C. and W. Easterling. 2001. Adaptation to climatic variability and change in the US great plains: A multi-scaled analysis of Ricardian climate sensitivities. *Agriculture, Ecosystems and Environment* 85: 133–144.

Risbey, J., M. Kandlikar, H. Dowlatabadi, and D. Graetz. 1999. Scale, context, and decision making in agricultural adaptation to climate variability and change. *Mitigation and Adaptation Strategies for Global Change* 4:137–165.

Robertson, G., E. Paul, and R. Harwood. 2000. Greenhouse gases in intensive agriculture: Contributions of individual gases to the radiative forcing of the atmosphere. *Science* 289:1922–1925.

Rochester, A. 1940. *Why Farmers Are Poor* (1975 edition). New York: Arno Press.

Rosenzweig, C. 1990. Crop response to climate change in the southern Great Plains: A simulation study. *Professional Geographer* 42:20–39.

Rosenzweig, C., and M. Parry. 1994. Potential impacts of climate change on world food supply. *Nature* 367:133–138.

Schneider, S., W. Easterling, and L. Mearns. 2000. Adaptation: Sensitivity to natural variability, agent assumptions and dynamic climate changes. *Climatic Change* 45(1):203–221.

Smit, B., and O. Pilifosova. 2001. Adaptation in the context of equity and sustainable development. Chapter 18 in *IPCC Second Assessment Report*. Cambridge: Cambridge University Press.

Smit, B., and M. Skinner. 2002. Adaptation options in agriculture to climate change: A typology. *Mitigation and Adaptation Strategies for Global Change* 7: 85–114.

Smit, B., D. McNabb, and J. Smithers. 1996. Agricultural adaptation to climatic variation. *Climatic Change* 33:7–29.

Smit, B., R. Blain, and P. Keddie. 1997. Corn hybrid selection and climatic variability: Gambling with nature? *Canadian Geographer* 41:429–438.

Smit B., I. Burton, R. Klein, and J. Wandel. 2000. An anatomy of adaptation to climate change and variability. *Climatic Change* 45:223–251.

Smit, B., and J. Wandel. 2006. Adaptation, adaptive capacity and vulnerability. *Global Environmental Change* 16:282–292.

Smithers, J., and B. Smit. 1997. Agricultural system response to environmental stress. In *Agricultural Restructuring and Sustainability*, ed. by B. Ilbery, Q. Chiotti, and T. Rickard, 167–184. Wallingford, UK: CAB International.

Statscan 2001. Statistics Canada Census 2001. Available at www.statscan.ca.

Tol, R.S.J., S. Fankhauser, and J.B. Smith. 1998. The scope for adaptation to climate change: What can we learn from the impact literature? *Global Environmental Change* 8:109–123.

Wheaton, E.E., and D.C. McIver. 1999. A Framework and key questions for adapting to climate variability and change, *Mitigation and Adaptation Strategies for Global Change* 4: 215–225.

Yohe, G. 2000. Assessing the role of adaptation in evaluating vulnerability to climate change. *Climatic Change* 46:371–390.

PART THREE

Adaptation Tools
and Case Studies

The Contextual Importance of Uncertainty in Climate-Sensitive Decision Making

Toward an Integrative Decision-Centered Screening Tool

SUSANNE MOSER

As human-induced climate change is increasingly accepted as fact, and decision makers begin to grapple seriously with the policy and management implications, climatic changes have the potential to become relevant to decision making; but the challenges of effectively linking science to policymaking and management practice are real and difficult to overcome. While uncertainties in climate change projections matter in important ways to those who must design and decide on mitigation policies, this paper focuses on the relevance of uncertainty to resource and land management at various levels of governance that addresses adaptation. Clearly, decision makers in the Great Lakes region at local, state, and regional levels will face precisely such challenges. This then raises several important questions, including:

- In what ways can climate change science support adaptation decision making?
- When and to what extent does uncertainty in climate change projections matter to decision makers concerned with adaptation challenges?
- How do we frame—and contain—the amount and type of uncertainty analysis that matters for the decisions at hand?

- What do decision makers need to know about scientific uncertainties in order to account for them appropriately in their decisions?

These types of questions force us to link and integrate scientific advances forged on uncertainty assessments within weather forecasts, climate variability and change projections, and impact analyses with those made in the understanding of the role of science in practical decision- and policymaking. Echoing a vast body of literature[1] and experience, the ultimate goal of such an integrative effort is to ensure that scientific information effectively connects with the needs of decision makers as they begin to address adaptation questions.[2]

Importantly, this integrative work must *shift the focus to the decision maker,* the decision-making process, and the relevance of weather and climate information—and specifically the relevance of obtaining information about uncertainty in climate research. The objective then is to develop a systematic approach to determining where and when uncertainty matters: Where is the decision-making environment particularly sensitive to uncertainty in the information provided? Is it useful and necessary to produce an "end-to-end" characterization of uncertainty (e.g., from emissions scenarios to model uncertainties, to climate sensitivity, to climate impacts to vulnerability and adaptation or mitigation policy options), and if so, when should it be produced? And if not, what do decision makers need to constructively and appropriately take climate change into account in their decisions?

This chapter proposes such a systematic approach and illustrates it with examples relevant to the Great Lakes region. The approach has been tested already in a case study of adaptation decisions in coastal management in California, but additional testing in "real-world" contexts would help strengthen it and prove its broad utility. The following sections begin with a conceptual discussion of the usefulness and fit of scientific information in the decision-making process, present the basic premises and objectives of the proposed approach, and then lay it out in a way that is cognizant of the decision process and of the constraints that decision makers face. Along the way, the chapter offers examples to illustrate the meaning and application of the approach. Suggestions for testing the approach further are also made, before concluding with an appraisal of its potential usefulness and limits. The ultimate hope is that the proposed tool will give scientists and decision makers a procedure to identify those instances where (even uncertain) science can most effectively support decision making.

THE USEFULNESS OF SCIENTIFIC INFORMATION IN THE DECISION-MAKING PROCESS

Science that aims to support decision making must pass—at the very least—the usefulness test. Additional important criteria that need to be met in order for the science/decision-making interaction to work effectively have been identified. These criteria include salience or relevance, credibility (which Jones, Fischhoff, and Lach (1999) include indirectly in their usefulness criteria), legitimacy of process (Moser 1997; Gieryn 1999; Cash 1998; GEA Project 1997; Mitchell et al. 2006; Cash and Moser 2000), and efficacy (again, included under the rubric of "usefulness" by Jones et al. 1999). As many recent studies and reviews have found, there is no clear, natural, or easy fit between the world of research and that of decision making (figure 1a) (e.g., NRC 2009).[3] Instead, in most instances, that fit has to be actively created, or—as Sarewitz and Pielke (2007) put it—the supply and demand of science in decision making has to be carefully reconciled.

A number of studies have examined and illustrated the criteria that need to be met in order to have a science/decision-making fit. Particularly useful for the purposes here is that by Jones, Fischhoff, and Lach (1999) (figure 1b). The arrows go in both directions in the above depiction, suggesting that the initiative for seeking a better link between science and decision making can come from either end of the spectrum. In the case where decision makers actively seek scientific information, they have already passed the receptivity and maybe even the accessibility hurdles. In the case where scientists seek to offer their information to decision makers, there may already be at

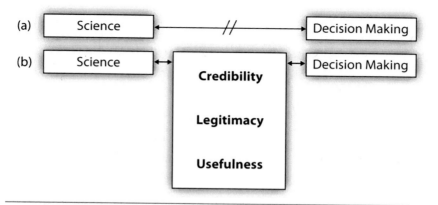

Figure 1. Linking science and decision making
Source: Expanding on Jones et al. (1999).

least a willingness to make that information relevant and compatible—even if the details have still to be worked out (further details on this below). From a pragmatic point of view, it would be naive, however, to assume this kind of mutual openness. Indeed, the fact that one or the other side is only minimally open to input from the other is frequently the biggest initial stumbling block between science and decision making (e.g., Lindblom 1980; Nutley, Walter, and Davies 2007).

A second critical issue to identify is where or when in the decision process science can insert itself, and what functions science can play at different stages of the decision process.[4] Typically, decision theorists identify a number of decision functions or stages in the decision process (e.g., Clark 2002). One of the more comprehensive depictions of this process, based on a rational approach to decision making, is given in figure 2.

Information needs differ significantly during different stages of this process (e.g., during the analytical intelligence gathering and the promotion stages). Similarly, the degree and nature of influence that scientific input has at different stages varies with the degree of usefulness (discussed above) and with the stage in the decision process at hand. In fact, science can influence the direction of that process, move it forward, or initiate another iteration of the decision-making process through its influence (Mitchell et al. 2006). To understand when and how that happens, it is often necessary to also understand the wider policy context and history of the decision-making process

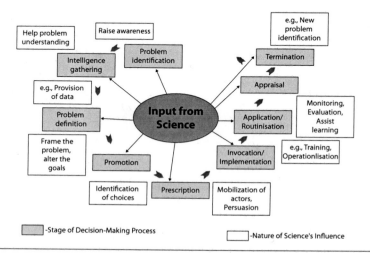

Figure 2. Scientific input at various stages of the decision process and the nature of influence
Source: Vogel et al. (2007).

itself (Herrick and Pendleton 2000; French and Geldermann 2005). Thus, this cycle of decision steps or decision functions should not be viewed as one in which steps follow each other linearly. Rather, "real-world" decision processes are iterative and messy, and science entering into that process will be drawn into that dynamic (IAI 2005). The decision process and the points of entry for science into it are delineated here for three reasons:

1. *Timing of scientific input is critical,* and scientists need to understand where in the process they enter the decision-making arena. The (unspoken) rules of their participation in the process may be quite different at different stages. Here, the meaning of "timing" refers to the stage in the decision-making process. Another aspect of timing relative to the decision calendar will be discussed below.

2. By highlighting the different nature of scientific input at various stages in the process, it becomes clear that *the nature and detail of scientific information required or acceptable will also vary.* As Herrick and Pendleton (2000, 358) maintained, "To suggest that predictive modeling may not be especially useful in some situation . . . is not to argue that scientific information will not be useful. Not all 'good science' is predictive."

3. *Scientific uncertainty* (and hence the need to characterize it with varying degrees of sophistication) *varies in its impact and importance across the different stages of the decision process.* In other words, scientific uncertainty is not uniformly important, but sensitive in its relevance to the specific decision context. For example, it may influence the way a problem is defined or the degree to which it is taken seriously. It may also help mobilize different sets of decision supporters or antagonists (actors, stakeholders) than if the problem was less uncertain. It may affect the set of choices available, or perceived as rational by a decision maker, and it may affect the weighing of options by the decision maker, and so on.

Finally, to put a finer point on the issue of usefulness, i.e., on finding the "right" science for the "right" entry point in the decision process, it is important to understand the management challenge from the perspective of the decision maker (e.g., Moser 2006; NRC 2009). Essentially, the issue at hand here is an institutional one, with historical roots in the evolution of management institutions, political boundaries, and other structural constraints on the decision-making process. Understanding this perspective is crucial in order to avoid some of the common types of frustration and misunderstandings between scientists and decision makers. For example, what may be a crucial logical or causal link between the climate and a given impact to the scientist may be completely irrelevant and uninteresting (beyond the sheer curiosity value) to the decision maker, if that particular impact falls outside his or her purview of authority.

To illustrate this point in more detail, it may be helpful to use a concrete example in which a scientist's and a decision maker's issue perception (and problem definition) significantly differ. A hypothetical climate-impacts assessor may, for example, study the impact of climate variability and change on the amount of thermally suitable habitat of commercially harvested whitefish. The scientific assessment suggests that suitable habitat may decline significantly, especially in shallow lakes, over the next 50 years (Kling et al. 2003; Mackey, this volume). This potentially highly relevant scientific information is offered to a busy decision maker at the state level in Michigan, who—after merely a glance—places the report onto the growing "read-(maybe)-later" pile.

Meanwhile, that same decision maker sits in his office and has to determine how to balance the competing needs and objectives for the particular streams and lakes in his jurisdiction over the coming year: the productivity of whitefish and other commercial and recreational fisheries, hydropower generation, endangered species protection, and maintenance of recreational opportunities. The issue of flood protection is also a consideration, but only tangential to his particular area of influence. It falls under someone else's decision authority. The varying competing objectives are discussed and decided upon in a watershed-wide working group, but some goals are also set in the state capitol or by Congress. The decision maker—in his given sphere of control—has to find a compromise between all these goals with a particular set of management options at his disposal, and not for the next 50 years, but for the coming season or year. Besides the objectives that have to be met, there are political, social, institutional, and economic considerations that also constrain or favor particular outcomes and management options— issues that would be fatal professionally to neglect.[5]

For the climate-impacts study to enter effectively into the decision challenge that this manager faces, the *scientific information has to link to the specific decision objectives* (outcomes, goals) *and/or the feasible management options* (choice set, levers) at hand for a given *decision timeframe,* i.e., the time scale over which today's decisions have implications, *at the right time.* Typically, decision makers and scientists—in collaboration—have to build the conceptual and data "bridge" between their different issue definitions and match the delivery of information with the decision calendar. In the example above, the link would have to be made between climate, lake water levels, water temperature and other habitat quality criteria, whitefish ecology, and fisheries economics. Several of these variables relate to the competing objectives the decision maker has to balance (e.g., lake levels vs. hydropower generation; water temperature, habitat quality, and fish ecology vs. endangered species protection; whitefish survival rate vs. fisheries productivity and economic impacts). Further scientific analysis could lay out under what climatic conditions the management objectives can no longer be met. Alternatively, further research could identify whether novel or altered management strategies could help meet the given objectives. In addition, these altered management strategies would have to fit into the decision calendar (timely information relative to the seasonal needs of the decision maker) (e.g., Pulwarty and Melis 2001; Sarewitz and Pielke 2007; McNie 2007; Tribbia and Moser 2008) and be acceptable to all stakeholders involved.

It is only then that theoretically relevant scientific information has truly met decision-making needs. It would have met the four conditions of usefulness mentioned above (see figure 1) (adapted from Jones, Fischhoff, and Lach [1999]):

$$\text{Usefulness} = f(Rv, C, A, Rc)$$

Rv: Relevance
- Does the research time scale match the decision-making time scale?
- Can climate variables affect the parameters under control of the decision maker?
- Can climate variables affect the decision outcomes?

C: Compatibility
- Is climate science output compatible with the form needed for decision making?
- Does climate science output feed into existing decision models or procedures?
- Can existing decision models accept probabilistic information as input?

A: Accessibility
- Where in the decision process can this climate science output enter?
- Can the decision-maker get the information?

Rc: Receptivity
- Is the climate science perceived as a credible and legitimate input to the decision process?
- Is the decision maker (and other relevant stakeholders) willing to use climate-science output?
- Do decision makers consider climate information worth knowing?

Based on the above discussion of the science/decision-making linkage, it is now possible to lay out the premises and objectives of a screening tool that identifies the usefulness of climate information and the need for information about uncertainty. The basic argument running through this paper is that where and whether climate and uncertainty information matter is an empirical question, rather than one that can be answered ex ante (see also Moss 2007).

PREMISES AND OBJECTIVES

The approach proposed here springs from a number of premises—each flowing from the understanding of the science/policy interface described above. Ideally, it will satisfy these requirements and simultaneously meet a series of objectives. In cases where it does not, the hope is that it can—with relatively little effort on a case-by-case basis—be adapted to meet them. The overarching goal, as stated above, is to develop a widely applicable tool that links the scientific analysis with information use and helps to identify those instances where uncertainty needs to be assessed, characterized, and communicated to the decision maker.

Premises

1. The approach must *place the decision maker and the real-world process of making a decision at the center.* Differently put, scientific products and information must fit into the actual decision-making process rather than into a theoretical model of decision making, or simply serve to advance scientific knowledge (NRC 2005, 2009; NRC Roundtable on Science and Technology for Sustainability 2005) (see also Scheraga, this volume).

2. The approach *does not—a priori—lend primacy to science and scientific information in the decision-making process over other decision inputs, but it does assume that credible, relevant, and accessible scientific information can be an important input* in many decisions. This importance is elevated to the extent that decisions affect complex systems, span longer time horizons, need to address significant risks and uncertainties, and pursue multiple objectives.

3. The approach *does not assume a particular normative approach to decision making under uncertainty* (e.g., a "wait and see" approach that favors delaying action in the face of uncertainty, or a precautionary approach that favors

preventive action in the face of uncertainty). Instead, it assumes that value judgments of this sort are made throughout the decision-making process, and that a well-informed decision process would benefit from a better understanding of the risks, uncertainties, complete unknowns, and the degree of confidence scientists place in particular climate information.

4. The approach *does not—a priori—favor a "top-down"* (global climate model to local impacts) *assessment approach, nor does it alternatively favor a "bottom-up"* (vulnerability-focused) *assessment approach* (Dessai and Hulme 2003). Instead, these two approaches are considered complementary, producing different scientific information, and implying a different decision focus or purpose, with implications and usefulness to the decision maker varying accordingly.

Objectives

1. The approach *should work for all kinds of weather and climate-sensitive decisions,* rather than be narrowly defined to work only for questions of climate variability, or only for questions of long-term climate change.
2. The approach *should be applicable in a variety of decision-making contexts.* This may include decision contexts in various natural-resource or hazard management situations (e.g., water, agriculture, forest, or coastal management)—that is, contexts where ongoing resource management and potential adaptation decisions will have to be made. In principle, however, the approach should also work in contexts where decisions may be focused on mitigation efforts (e.g., in the energy or transportation sectors).
3. The approach *should also work for a range of decision makers,* be they in the private sector, public sector, or in mixed settings. Typically, multiple decision makers are involved, but there may be a lead or coordinating authority.
4. The approach *should be applicable at a variety of scales.* Many climate-sensitive resource-related (adaptation) decisions are made at state, regional, and local levels, but may also involve higher levels of governance. Mitigation decisions (or planning/development decisions that affect greenhouse gas emissions) tend to be made at international, national, or state levels, but frequently do involve regional and local levels.

The following section introduces an approach that aims to meet the premises and objectives discussed above while being cognizant of the realities of decision making. Subsequent sections illustrate a case example of how this tool can be applied. Further research will be necessary to empirically test and refine it, and to determine the larger significance of this approach.

DUST—DECISION UNCERTAINTY SCREENING TOOL

To systematize the identification of climate and uncertainty information needs in the course of decision making, it is useful to classify decision situations. Different types of decisions pose different scientific challenges, and hence require different approaches to characterizing scientific uncertainty. Many attempts at classifying decisions exist, using, for example, the substantive decision context (e.g., food production, energy distribution, water resource management), the scope or magnitude of the decision (e.g., measured in affected dollar value), the time horizon of the decision (e.g., short- versus long-term), or the type of decision in the management context (e.g., operational, investment, design, or planning decisions) as the underlying principle of distinction (e.g., NRC 1981; Clark 2002; Sarewitz, Pielke Jr., and Byerly Jr. 2000).

For DUST, decisions are primarily categorized using the fundamental elements that are common to all decisions. The basic "building blocks" of decisions are:

- The *present conditions* (P) (state variables) defining the baseline or the perceived problem;
- The *objective(s)* (O) or goals of a decision (sometimes with specified *criteria* (c) that would satisfy these objectives);
- The *choice set* (C) or management options (levers or control variables) available to achieve the objective(s) (sometimes with explicit, but often only implicit *attributes or preferences* (a) attached to each choice);
- Decision *constraints* (X) (such as social, technical, economic, or political factors that arise in the context in which the decision is being made); and
- *Externalities* (E) (known or unexpected impacts that arise from the decision that were not explicitly included in the weighing of the decision).

Underlying all of these basic building blocks is a specific problem definition (for further discussion, see the section on "Usefulness of Scientific Information in the Decision-Making Process" above), which as a frame for the decision problem mobilizes certain perceptions and conceptualizations of the problem, potential solutions, and feasible means to resolve it. Using these decision elements, two fundamentally different types of decisions are derived—optimization and evaluation decisions. Each is described in turn below.

Optimization Decisions

The basic question the decision maker asks in this type of decision is: What decision (i.e., what strategies or choices) will produce the desired outcome? Or more colloquially, what path shall I pick (choices) to get to Rome (outcome)? This decision problem has also been described as "learn now, then act" (Kann and Weyant 2000). More formally, this type of decision can be depicted as:

$$O_c \ (+ \ E) \ \rightarrow \ C_a \ f(P, \ X)?$$

Given the present situation (P) and certain constraints (X), which choices (C) with specified attributes (a) will optimally combine to produce the desired (c) outcome (O) (and minimize the creation of unacceptable externalities (E))?

Optimization decisions have a specified outcome in mind and involve choosing among a set of choices or strategies to achieve it. Outcomes are typically very specific, but in general can be described as achievement of a positive outcome, avoidance of a negative outcome, or compliance with a required outcome (which essentially singles out one of the two other possible objectives).

Note that the term "optimization" does not imply an unrealistic assumption about utility maximizing or other forms of finding an "objectively ideal" solution to a problem. Rather, most decision makers will "satisfice," i.e., choose from a limited set of options with bounded rationality. "Optimization" simply refers to the fundamental type of decision problem. (See also endnote 7, and Marx and Weber, this volume).

Examples of climate-sensitive optimization decision problems
1. What is the best way (C) to protect a home from devastating flood loss (O)? Choices for the homeowner may include elevating or structurally reinforcing the home, removing heaters from the flood-prone ground floor, buying (more) flood insurance, doing nothing and

hoping that there will be no major flood, etc. Choices at the local or state government level may include changes in dam operations, building of flood retention basins, strict implementation of building codes, removal of houses out of a flood plain, better flood insurance coverage, etc.

2. *To cost-effectively minimize (O) the number and severity of occasions when the sewer system in the southern Great Lakes region is overwhelmed by storm runoff and regular discharge of affluent, where and how quickly do existing sewage pipes have to be replaced with bigger ones (C)?* In this instance, the choice set is already reduced to an engineering solution (land-use and management changes are not considered here), but the question is where to focus first, and what the replacement pipe size should be to minimize both costs and spillage events. (For a detailed description of sewage and stormwater runoff challenges with climate change, see Scheraga, this volume)

3. *By 2100, CO_2 concentration in the atmosphere must be 450ppm or less (O). How can the global community get there (C)?* Choices at any level of government may include various combinations of carbon taxes, regulations, trading and incentive programs, technology investments, and so on. In order to minimize the cost of achieving the ambitious goal, approaches may be combined in a particular way, while additional considerations—such as equity, feasibility, political acceptability, etc.— may lead to a different "optimal" combination of approaches.

Evaluation Decisions

The basic question the decision maker asks in this type of decision is: What outcome does a given (set of) decision(s) have? More colloquially put, a decision maker may ask, I wonder where all these different paths lead? (Also described as an "act now, then learn" problem in the words of Kann and Weyant [2000].) More formally expressed, this type of decision can be depicted as:

$$C_a \rightarrow O_c \text{ } (+E) \text{ } f(P, X)?$$

Given the present situation (P) and certain constraints (X), what outcome (O) with certain desired criteria (c) will result from selecting particular choices (C) with specified attributes (a), and what externalities (E) may they produce?

Evaluation decisions start out from a set of policy options or strategies and assess the potential outcomes and tradeoffs among these choices.[6] Evaluation decisions may ask:

- What are the advantages and disadvantages of each of the choices in my set?
- Which single choice from my set is the best (or worst) with regard to a specified attribute or outcome criterion?
- What combination of choices maximizes the benefits and opportunities (or minimizes the risks, costs, or other negative consequences) in the aggregate?
- What combination of choices distributes the risks, costs, and benefits most equitably (or politically most defensibly)?
- Which (combination of) choice(s) retains the greatest amount of future flexibility, which involves irreversibility?

Examples of climate-sensitive evaluation decision problems

1. To ensure the survival of endangered species, such as Michigan's Kirtland's warbler (O), should conservation planners protect habitat (C1), protect the jack pine on which the bird depends (C2), or try to physically relocate the warbler to habitats with similar soil, vegetation, and temperature conditions further north (C3)? This decision is between different management options to achieve one desirable goal. (For more discussion of challenges faced by plant and wildlife species with a rapidly changing climate, see Root and Hall, this volume.)

2. What is the cost and benefit (O) of increasing irrigation of a farmer's crops (C) if temperature and precipitation changes manifest as currently projected? This decision does not start out from a desired agricultural yield, but wants to know the impact of a particular management approach and the net income of the farmer. (For more detailed discussion of agriculture and climate change, see Easterling et al., this volume.)

3. Given projected changes in streamflows and lake levels, how would a certain change in the controlled release of water (C) at the Robert Moses Niagara Power Plant, downstream of the Niagara Falls, affect instream water flows over the course of the year (important for aquatic life) (O1) and overall energy production (O2)? This is a typical trade-off question that requires management strategies for multiple objectives to be balanced in new ways.

To determine the acceptability of the final decision, decision makers in both cases, i.e., for either type of decision, must take into account the criteria by which outcomes should be measured, the attributes that characterize the available choices, and any other constraints and externalities affecting the choice.

The simple distinction between the two different types of decision challenges neglects the fact that knowledge about any one aspect in these decisions varies from relatively certain to uncertain, to deeply uncertain, to completely unknown. This fact, while not novel to decision making (albeit pressing in the case of long-term climate change), suggests very different analytical approaches from a scientific perspective (Morgan and Henrion 1990; Lempert, Popper, and Bankes 2003) (see also the discussion in Easterling et al., this volume). Where deep uncertainty and ignorance persist, subjective expert judgment becomes a critical and necessary input. In communicating uncertainty to decision makers (see below), this needs to be made transparent.

FURTHER NECESSARY DISTINCTIONS AMONG DIFFERENT TYPES OF DECISIONS

Considering the ultimate interest in what kind of climate information is needed, to what extent uncertainty matters, and if so, how it should best be assessed and characterized, further distinctions need to be made among the already identified basic types of decisions. The first of these relates to the decision time horizon. Does the decision pertain to the near future (e.g., an extreme weather event, a growing season, or a few years), or does it reach into the far future (as perceived by the decision maker), such as in siting decisions (e.g., 5, 10, 25, 50, or 70 years)? Another dimension relates to the number of opportunities to revisit a decision, or—differently put—the number of iterations in the decision-making process. Is the decision a one-time decision, or are there opportunities for sequential decisions (including updates of the data informing the decision, and learning) over time? The implications of these distinctions for the demands on scientific information are made clear in the following examples:

1. *Near-term/single decision moment:* A farmer in Illinois has to choose the set of crops to plant for the coming year. Long-range seasonal forecasts suggest it is likely to be a wetter-than-normal year. Spring weather has already been

very wet and caused some delays in planting dates. But if moisture levels are maintained without being excessive during key times of the growth cycle, crop yields could be very good. Once the seed is in the ground, that farmer will have to deal with the consequences of his choice, no matter how the rest of the year turns out in terms of weather. If it continues to be wetter than usual, he may encounter harvest losses; if it becomes drier than usual, he may have to absorb the cost of irrigation or the loss in yield. Or he may just turn out lucky (see the chapter by Easterling et al., this volume, for a detailed discussion of uncertainty in agricultural decision making). This is an evaluation problem.

2. *Near-term/multiple decision moments:* Over the course of the year, a water resource manager in Michigan (e.g., an operator of one of the 13 remaining hydropower facilities along the Manistee, Muskegon, Au Sable, Grand, and Kalamazoo Rivers) has to make numerous decisions adjusting the water volume/level in the reservoir. Each time updated weather and climatic forecasts are available, some limited adjustments can be made based on the newer information as to how much water should be stored or released. Given the seasonality of precipitation, runoff, air and stream temperature, and changes in electricity demand by consumers, as well as the water needs of aquatic life, however, some of these decision points involve irreversible decisions (for that storm event or season). The water released at present to protect against possible major runoff and flooding episodes, or to provide cooler and more water to fish, cannot be returned for storage (and released at a later time) for electricity production during times when the water and power demands peak. This is an optimization problem.

3. *Long-term/single decision moment:* For years, a small community on the shores of Lake Superior may have dealt with lake level fluctuations and shoreline erosion by demanding that homes be set back a reasonable distance from the shoreline and moved back if and when erosion threatens to undermine a building, but may have also allowed decks and piers to be built further out into the lake when lake levels were lower to permit water access for recreational boating. Current climate change projections suggest that lake levels will fall as climate warms, potentially encouraging greater shoreline development, and development closer to the water's edge. Despite the projected long-term trend of lake-level fall, historical experience suggests that lake levels vary, and scientists emphasize that they will continue to do so in the future. Exposed lake sediments may be toxic, and water-quality concerns suggest that limiting development may continue to be a good idea (see the detailed discussion of shoreline-management challenges with climate change discussed by Mackey, this volume). What should the community do? What

is a reasonable course of action? How should long-term trends, the range of risks, and the societal benefits of development be weighed against each other? This is an evaluation problem.

4. *Long-term/multiple decision moments:* For example, in light of credible new projections of climate change, a public health official in Chicago may be charged with designing a heat-emergency management system for the metropolitan area. She is considering a program involving multiple stakeholders (from all levels of government), and with multiple elements or management options that will be triggered by different levels of criticality. Criticality is defined by a combination of meteorological and social conditions. Over the years, the new system is being put in place and tested as heat waves hit the city. In numerous places, the system works smoothly; in others it fails. While the process is time-consuming and requires considerable staff resources, and regrettable shortfalls affecting individuals' health are repeatedly being experienced, lessons are being learned and incorporated in subsequent management of heat emergency situations. As heat waves get worse with climate change, the city is nonetheless adequately prepared. This is an optimization problem.

Essentially then, decisions can be categorized into a three-dimensional space that spans between decision types, decision time horizons, and decision opportunities (see figure 3).

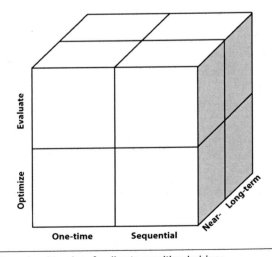

Figure 3. Three-dimensional typology for climate-sensitive decisions
Source: Created by author.

Based on this three-dimensional decision typology (loosely derived from Kann and Weyant [2000]), we can now identify different types of policy and uncertainty analyses that could be conducted.

UNCERTAINTY ANALYSES FOR DIFFERENT DECISION PROBLEMS

In general, sequential decision making under uncertainty can make use of uncertainty-assessment approaches that determine optimal policies at different points in time. This allows updates of all critical aspects of the decision: input/state variables, data informing the decision, or the decision set (control variables) and any of the criteria used to evaluate outcomes. In short, sequential decision making allows learning, and in some instances can be less risky, at least for those decisions that can be made incrementally (see also NRC 2009).

Where a decision is made only once over a given decision time frame, uncertainty grows with the length of the time frame for all aspects of the decision. The only exception is information about current conditions and the currently available set of decision options. Such decision situations do not allow learning and adjustment over time. They only allow finding out whether a decision was—based on the best judgment at the time—a good or a bad one, thus permitting some learning for similarly challenging future decisions. In these types of one-time decisions, then, whatever uncertainty is present at the time of decision making will propagate through the optimization or evaluation process, hence the use of so-called uncertainty propagation models in policy analysis. The following tabular overview of approaches to policy analysis hinges on the type of problem a decision maker faces (figure 4; for mathematical formulations of these different types of policy analyses, see e.g., Kann and Weyant 2000; Morgan and Henrion 1990).

Once the basic type of analysis required has been identified, the next step is to formulate—formally or informally—the decision problem using the basic decision elements. This problem formulation helps identify what is understood with high confidence, and what is more uncertain, entirely unknown, or can only be subjectively evaluated. These distinctions will then enable formal uncertainty analyses of the involved data, model parameters, and underlying model structures (the three basic types of uncertainty commonly discussed in the literature). Such quantitative approaches have been described in great detail in classic treatises such as Morgan and Henrion

Types of Decision	Policy Analyses	Remarks
One-time, near-term optimization	Optimization with resolved (known) uncertainty	Essentially a special case of stochastic dynamic optimization
One-time, long-term optimization	Finite-horizon stochastic optimization	
Sequential, near-term optimization	Infinite-horizon (dynamic) stochastic optimization	May be too computationally demanding
Sequential, long-term optimization	Infinite-horizon (dynamic) stochastic optimization	Quite computationally demanding
One-time, near-term evaluation	Single-period (but multiple policies) decision analysis; Single-policy uncertainty analysis	
One-time, long-term evaluation	Single-period (but multiple policies) decision analysis; Single-policy uncertainty analysis	
Sequential, near-term evaluation	Multi-period decision analysis	
Sequential, long-term evaluation	Multi-period decision analysis	

Figure 4. Types of decisions and relevant policy analyses
Sources: Based on Kann and Weyant (2000) and Morgan and Henrion (1990).

1990; Cullen and Frey 1998; Edwards, Miles, and von Winterfeldt 2007. Figure 5 below provides a suggestive overview of the types of analyses that could be done and what type of information they would yield.

COMMUNICATING UNCERTAIN RESULTS

The science/decision-making interaction does not end with the formal decision and uncertainty analysis. The obtained results must be presented in understandable form back to the decision maker (e.g., Webster 2003; Patt

Types of Analysis	What Can Be Learned
Exploratory modeling/Computer-assisted reasoning	Reveals model-based uncertainties and unknowns; used to explore plausible futures where little is known about them
Multi-model comparison	Reveals model-based uncertainties; important when model structures are less well known
Sensitivity analysis	Reveals the impact of varying model inputs (through single or joint variation); important when model structure is well known
Multi-scenario comparison	Reveals the impacts of different assumptions about the world (can be understood as a subset of the sensitivity analysis)
Propagation of uncertainty in input variables through a deterministic (or stochastic) model (e.g., use of decision trees, numerical simulation techniques or expert solicitation to develop plausible distributions for input variables)	Reveals the spread (frequency and/or probability) of outcomes due to this uncertainty in the input variable
Value of Information, Value of Uncertainty techniques	Reveal the impact of having perfect knowledge or having knowledge about uncertainty on a specified outcome
Model validation/comparison against empirical data or analogues in time or space	Suggests a level of confidence one can have in model results

Figure 5. Information gleaned from various types of uncertainty analyses
Sources: Based on Kann and Weyant (2000), Morgan and Henrion (1990), Lempert (2002), and Lempert, Popper, and Bankes (2002, 2003).

and Dessai 2005; Dabelko 2005; Moser 2006; Moss 2007; Patt 2007; Morgan et al. 2009). Critical considerations here include the following:

- Some decision makers are increasingly familiar with, or even trained in, uncertainty analysis, but this should not be assumed as the norm.
- Results from an uncertainty analysis can—in principle—be presented numerically, graphically, or simply in words. Which of these forms works best with the specific decision maker at hand must be explored in the specific

situation. Descriptive terms used to distinguish levels of uncertainty should be consistently defined and used (NRC 2006; Moss and Schneider 2000; Morgan et al. 2009).

- Whatever form of communication is chosen, the link must be made back to the decision problem (objectives and decision-choice sets) the manager faces in light of the stage of the decision process (Moser 2006).

- Experience reported by scientists and decision makers reiterates over and again that it may not be enough to quantify the degree of uncertainty; frequently, decision makers need or want transparency (Herrick and Pendleton 2000), especially when subjective expert judgment is involved. They want to understand the reasons for the uncertainty (e.g., is it due to natural variability, lack of understanding of the processes producing certain impacts, unpredictable changes in technology or society more generally, or underlying assumptions—such as discount rates—of the model?). This deeper understanding helps to increase the likelihood that communicated uncertainty is actually understood adequately (Pielke 2001; NRC 1989, 1996; Morgan et al. 2009). It also will help decision makers in their assessment of the climate information, and in turn to explain certain decision choices to other stakeholders. In any multi-stakeholder/multi-decision-maker situation, such explicit communication is particularly critical (e.g., Demeritt and Langdon 2004).

PUTTING IT ALL TOGETHER

The Decision Uncertainty Screening Tool (DUST) proposed here is essentially a step-by-step process of identifying where and how climate science could inform the decision process, getting successively more specific, until the needed uncertainty analyses are identified and carried out, and the results are communicated back to the decision maker. As a systematic, step-by-step procedure, it is likely to be much more linear and clear than the actual process may turn out to be. But for clarity's sake, the seven steps are summarized together with a reminder about the purpose of each step. This summary then links together the range of aspects discussed above: usefulness of science in the decision process, decision typologies, policy analyses, uncertainty assessments, and communication.

Step 1: Identify the Stage in the Decision Process Where Science Would Enter

The usefulness of science varies by stage in the decision-making process. The first step in DUST is to identify the stage of the process (figure 1), so that the most useful input at that particular time and in the specific context of the decision-making process can be determined. The goal is the identification of places where science would enter the decision-making process, and what the nature of such input would be.

Step 2: Explore Whether Scientific Input Would Be Truly Useful

Through direct interaction between scientist and practitioner, it can be determined in more detail what type of scientific input would be most useful. The goal is to obtain a clearer understanding of the decision maker's receptivity; the venues, formats, and specific data needs; as well as any relevant timing issues that would make information useful.

Step 3: Identify the Specific Decision Challenge

Step 3 in DUST is to identify the particular decision that practitioners face, i.e., whether it is an optimization or evaluation type of decision. As a first approximation, scientists ascertain the specific decision challenge at hand: what are the decision objectives and choices to which scientific information has to be linked?

Step 4: Identify the Type of Decision Problem the Decision Maker Faces

In step 4 of DUST, additional information is obtained from the decision maker to more clearly specify the type of decision that is being faced: what is the decision time horizon, and how many opportunities are there to revisit the decision? The answers to these questions help identify the appropriate type of policy/decision analysis to be conducted.

Step 5: Identify Necessary Uncertainty Analyses

The type of uncertain decision that practitioners face (identified in steps 3 and 4) help identify the relevant uncertainty analysis. In step 5, then, the appropriate uncertainty assessment is matched with decision-relevant variables and objectives. Much time and energy can be saved by focusing only on those analyses that can truly make a difference for the decision at hand.

Step 6: Conduct Identified Uncertainty Analyses

Finally, step 6 is to actually conduct the relevant and necessary uncertainty analysis to answer the remaining decision-relevant question: how does uncertainty in specified aspects of the problem affect the decision?

Step 7: Communicate Uncertainties Back to the Decision Maker

Once the uncertainty analysis has been completed, the results have to be communicated back to the decision maker in effective ways. "Effectiveness" does not just mean quantitative accuracy, but also using verbal and graphic formats that are understandable and meaningful to the decision maker, and pertinent to the decision at hand. The goal is to provide effective feedback to the decision maker, linking analysis to his or her specific decision problem.

TOWARD GREATER REFINEMENT OF DUST

DUST has been applied in a case study of adaptation to climate change by California coastal managers (Moser 2005). The extent of such local adaptation at the time (2005–06) was so limited that detailed uncertainty analysis was not yet necessary. Following the step-by-step approach revealed, however, what types of information California coastal managers would need, including what they would like to know about the types, nature, and degree of uncertainty (Tribbia and Moser 2008). These insights can better inform what type of science is being conducted, and has led to several changes in information and training services provided by such agencies as NOAA and the San Francisco Bay Conservation and Development Commission.

Further testing of cases that progress through the entire suite of steps involved in DUST would be highly desirable. Ideally, the testing of DUST would involve both scientists and decision makers. Pragmatically, this may be the most important and challenging thing to achieve. As mentioned above, to get scientists and decision makers to collaborate can be the most difficult step of all (NRC 2009).

Analysis of empirical test cases would not involve a full uncertainty assessment (steps 6 and 7, above) for all identified instances where it might be useful and scientifically feasible. Instead, the testing would be conducted in a joint working session in which

- decision makers assess whether their decision process is captured adequately in concept;
- decision makers offer empirical descriptive detail of their specific decision, involved players, relevant criteria and objectives, and the decision process;
- decision makers identify their climate-information needs, or at least identify climate-sensitive decision points (if necessary with the help of scientists);
- scientists explain in general terms what type of climate information they could offer—e.g., global model outputs, regionally downscaled projections, for certain climate variables (of course, any one scientist cannot provide all of this information, but within the project, the relevant expertise may be available);
- scientists explain in general terms the current level of understanding and confidence in the information they could provide (e.g., what factors principally determine the output, how well processes are understood, how well processes or outcomes can be predicted, what cannot be known and why);
- decision makers and scientists together determine in iterative fashion what information about climate and about uncertainty is needed specifically, in what form, and when and how frequently, as well as how uncertainties would need to be described (in words, numbers, or graphically) to be understandable and useful to decision makers.

This process should be recorded by the participants or a recorder. The results of working through the DUST model are likely to confirm some aspects of it and suggest modifications of others. Integrating test results into a revised DUST model is also likely to be iterative as multiple test cases may suggest alternatives or confirm consistent modifications. The testing may also result in the development of different versions of the screening tool for different decision situations. In short, the question of evaluating the test results of the DUST model should not simply result in a conclusion such as "it is useful or it is not," but instead result in one or more revised versions of the screening tool to help scientists and decision makers identify useful entry points for scientific information in specific decision situations (see also NRC 2009).

An additional benefit of the testing may be the identification of ways to simplify or streamline the DUST model. The final model, however, should be—in Einstein's words—"as simple as possible, but not simpler."

WHAT DUST IS NOT

Before concluding with a reiteration of the potential value of the DUST model, it may be useful to reiterate what DUST is not. Clearly, DUST is . . .

- **Not an Integrated Assessment Tool:** Integrated assessments are models that quantitatively link inputs (drivers) and outputs (impacts) from a range of complex submodels (e.g., climate, natural resources, economic sectors). Frequently they are used to assess the consequences of certain policies or drivers of change on sectors of interest. Many integrated assessors have urged that uncertainty analysis be an integral part of such complex modeling. DUST can be viewed as a support tool to integrated assessments.
- **Not a Policy Analysis Tool:** Policy analysis employs tools such as cost-benefit analysis, which in turn rests on a range of theoretical underpinnings, such as utility theory, contingent valuation, or statistical decision theory, to identify defined optimal strategies or outcomes. While DUST starts out from a distinction of fundamentally different decision types that link outcomes with means, it only uses this categorization to aid the identification of appropriate uncertainty-analysis approaches.
- **Not a Decision-Making Tool:** As a tool that can help identify information needs, in particular about uncertainty of climate science, it can be viewed as a decision *support* tool. By itself, however, it will not help identify preferred solutions, but only identify information that may help decide between potential solutions. An effort was made to avoid theory-prescribed, unrealistic assumptions about decision situations, decision makers, or their objectives.
- **Not an Uncertainty Assessment Methodology:** DUST simply aims to identify when and where what types of existing uncertainty analyses should be conducted in particular decision situations. As such, it parallels the approach described in Kann and Weyant (2000).

While DUST is theoretically informed (e.g., by decision-theoretical ideas such as bounded rationality, and economic concepts such as satisficing), it aims to reflect the "messiness" of empirical reality more so than any one particular theoretical understanding of the decision-making process (see also Marx and Weber, this volume). As such, it also tries to avoid many of the pitfalls of conventional policy analysis and decision-support tools (Morgan et al. 1999). For example, DUST places the decision maker and his/her challenges in the center. This should not be read as a conceptualization of the decision-making situation as consisting of only one decision maker—clearly an unrealistic situation in most contexts. Instead, there will almost always be several decision makers with different (even competing) information needs, objectives, and constraints. But, to match scientific information effectively with the decision makers' information need, the fit must be individualized (or at least match the needs of groups of similar decision makers).

DUST is also designed to be flexible enough to transfer from context to context. It does not assume a static set of values, objectives, or an unrealistic set of managerial choices. It simply creates a systematic framework in which the actual set constituting the decision situation must be considered.

Maybe most importantly, the approach does not assume that climate information must be used in a particular situation, even if theoretically there exists a logical link between climate and what is being managed. It simply offers a systematic method to find out whether climate and uncertainty information could be useful to the decision maker (see the valuable discussion in Dessai et al. 2009). The uncertainty-analysis tools identified through DUST may reveal, for example, that (better) climate information would not substantially affect or improve a given decision. This by itself would be a valuable finding that could initiate additional research or changes in the decision-making process (for a case study of this situation, see Jones, Fischhoff, and Lach 1999).

A final limit of the approach offered here is that DUST does not address the many decision-external factors, such as institutional barriers, personal competition or limitations, lack of social acceptability, or lack of resources that frequently constrain the use of scientific information in decision making (Moser 1998; Moser 2009). Scientists, however, should be cognizant of such barriers and uncertainties in the human dimensions of the decision-making process. Appropriate framing of the decision problem may help overcome some of them.

POTENTIAL BENEFITS AND IMPACTS OF THIS APPROACH

Testing and refining DUST to enhance its usefulness is a crucial step toward integrating geophysical climate-modeling work, the impacts analyses, decision science, and science policy, as well as the world of practical decision making. If the tool proves useful, it would help the research community come closer to fulfilling its promise of producing assessment science that closes the science/society gap, thus becoming truly useful to decision makers.

From the outset, the screening tool is designed to be "transferable" to a range of decision-making contexts (see Premises and Objectives above). Further vetting among experts and testing in a range of empirical settings will make DUST a useful approach that is available for wider application by the scientific and practitioner community. It offers itself particularly for use in existing or future scientist/decision-maker collaborations (e.g., applied science problems, stakeholder-informed assessment processes, ongoing integrated climate-assessment efforts). It may also be of educational value to beginning scholars or students entering into applied science fields where a better understanding of the science/decision-making interaction would be particularly valuable.

If adopted for wider use, the impact of the DUST tool could go even further. Ideally, the proposed approach can serve multiple purposes:

1. From a science perspective, the approach may serve as a *streamlining and prioritization function for uncertainty assessment*. Ideally, it will help to systematically identify where, when, and with what scientific methods uncertainty should be assessed, to whom this uncertainty needs to be communicated, and what forms of characterization and explanation would be most useful to the decision maker.
2. From a decision-making perspective, the approach may lead to *greater transparency and awareness*.[7]
3. The approach may also serve as an *educational function for scientists* about the real-world decision-making process and the use and usefulness of scientific information in that process.
4. In a complementary fashion, the approach may also serve as an *educational function for decision makers* about the relevance of climate information to their decisions, about the state of knowledge of climate-change science, and about the degree of certainty and confidence scientists have in different aspects of the problem.

5. In accomplishing (3) and (4), the approach could also serve the function of a *boundary object*—i.e., a tangible product or tool around which scientists and decision makers can interact, learn from each other, fine-tune products, and build mutual trust and understanding, but also maintain the necessary boundary between science and decision making. As such, interaction around boundary objects can help ensure credibility and legitimacy while enhancing the relevance of scientific information to decision makers. It is easy to envision, for example, that only the early steps of the DUST tool are implemented, which would serve the boundary object and educational functions, and thus enhance the science/decision-making interface.

Thus, even a partial fulfillment of these envisioned and intended outcomes would be a major step forward in not only improving the usefulness of climate science for decision making, but, in fact, in elevating the public discussion about decision making in the face of uncertainty.

NOTES

1. For example, Baeckstrand 2002; Cash et al. 2003; Dresler and Schaefer 1998; Glasser 1995; Jones, Fischhoff, and Lach 1999; Malone and Yohe 2002; National Council for Science and the Environment 2000; Pielke Jr. 1997; Pielke Jr. and Conant 2003; Pulwarty and Melis 2001; Steel et al. 2004; Wynne 1992; Jacobs, Garfin, and Lenart 2005; van Kerkhoff 2005; Lemos and Morehouse 2005; van Kerkhoff and Lebel 2006; Karl, Susskind and Wallace 2007; and a synthesis of the literature relevant to climate-related decision support by National Research Council 2009).

2. The term "decision maker" is used here as shorthand to mean a wide range of individuals at different levels of governance deciding whether or not to take a certain course of action in any given climate-sensitive sector: public officials setting policies at federal, regional, state, or local levels; private-sector business managers helping to determine the business strategy of their company; private-sector or public-agency officials deciding about resource allocations; individuals determining their personal/private business operations (such as choosing crops, planting and harvesting dates) or choosing residential locations; and so on.

3. For example, Healey and Hennessey 1994; Healey and Ascher 1995; Jones, Fischhoff, and Lach 1999; Pulwarty 2003; Pulwarty and Melis 2001; Alcamo, Kreileman, and Leemans 1996; Baeckstrand 2002; Boesch 1999; Bradshaw

and Borchers 2000; Cash et al. 2003; Cashmore 2004; Catizzone 2000; Cortner 2000; Dresler and Schaefer 1998; Fabbri 1998; Glasser 1995; Gieryn 1999, 1995; James 1999; Korfmacher 1998; Malone and Yohe 2002; NRC Commission on Geosciences 1994; Pielke Jr. 1997; Pielke Jr. and Conant 2003; Swets, Dawes, and Monahan 2000; Moser 1998; Moser, Cash, and Clark 1998; NRC 1981; Hall and Lobina 2004).

4. For the purposes here, I will forgo an explicit treatment of the ways in which decision making can inform science at each of these stages. Fuller discussions of the processes in which "stakeholders" contribute to the development of scientific research agendas and assessment processes can be found elsewhere (e.g., http://www.harvard.edu/gea/). However, ideally, a back-and-forth iterative process between scientists and decision makers should evolve as a result of using the approach proposed here.

5. For an empirically documented case of just such differences between scientific input and decision-maker needs, see Moser (1997). Other match and mismatch cases are collected in Sarewitz, Pielke Jr., and Byerly Jr. (2000) and further discussed in Sarewitz and Pielke (2007).

6. A new type of methodology has been developed in recent years that may be seen as a hybrid of the optimization and evaluation decision types, depending on how the decision is formulated. This new method—called Robust Adaptive Planning—uses computer-assisted reasoning to examine management strategies that avoid major system failures, breakdowns, or surprises (Lempert, Popper, and Bankes 2002, 2003).

7. This may or may not always be a desirable outcome in the eyes of decision makers.

REFERENCES

Alcamo, J., E. Kreileman, and R. Leemans. 1996. Global models meet global policy: How can global and regional modellers connect with environmental policy makers? What has hindered them? What has helped? *Global Environmental Change* 6(4):255–259.

Baeckstrand, K. 2002. Civic science for sustainability: Reframing the role of scientific experts, policy-makers and citizens in environmental governance. Paper read at 2000 Berlin Conference on the Human Dimensions of Global Environmental Change: Knowledge for Sustainability Transition: The Challenge for Social Science, December 6–7, 2002, at Berlin, Germany.

Boesch, D.F. 1999. The role of science in ocean governance. *Ecological Economics* 31(2):189–198.

Bradshaw, G.A., and J.G. Borchers. 2000. Uncertainty as information: Narrowing the science-policy gap. *Conservation Ecology* 4(1):Art no. 1.

Cash, D. 1998. Assessing and Addressing Cross-scale Environmental Risks: Information and Decision Systems for the Management of the High Plains Aquifer. Diss., JFK School of Government, Harvard University.

Cash, D.W., W.C. Clark, F. Alcock et al. 2003. Knowledge systems for sustainable development. *PNAS* 100:8086–8091.

Cash, D.W., and S.C. Moser. 2000. Linking local and global scales: Designing dynamic assessment and management processes. *Global Environmental Change* 10:109–120.

Cashmore, M. 2004. The role of science in environmental impact assessment: Process and procedure versus purpose in the development of theory. *Environmental Impact Assessment Review* 24(4):403–426.

Catizzone, M. 2000. Building dialogue between scientists and policy makers! *IGBP Newsletter* 42:12–13.

Clark, T.W. 2002. *The Policy Process: A Practical Guide for Natural Resource Professionals*. New Haven, CT: Yale University Press.

Cortner, H.J. 2000. Making science relevant to environmental policy. *Environmental Science & Policy* 3(1):21–30.

Cullen, A.C., and H.C. Frey. 1998. *Probablistic Techniques in Exposure Assessment: A Handbook for Dealing with Variability and Uncertainty in Models and Inputs*. New York: Plenum Press.

Dabelko, G.D. 2005. Speaking their language: How to communicate better with policymakers and opinion shapers—and why academics should bother in the first place. *International Environmental Agreements* 5:381–386.

Demeritt, D., and D. Langdon. 2004. The UK Climate Change Programme and communication with local authorities. *Global Environmental Change* 14(4):325–336.

Dessai, S., and M. Hulme. 2003. Does climate policy need probabilities? In *Tyndall Centre for Climate Change Research Working Paper*. Norwich, UK: University of East Anglia, Tyndall Centre for Climate Change Research.

Dessai, S., M. Hulme, R. Lempert, and R. Pielke Jr. 2009. Do we need better predictions to adapt to a changing climate? *EOS Transactions of the AGU* 90(13):111–112.

Dresler, P., and M. Schaefer. 1998. Regional forums linking science and management. *Geotimes* 43(4):24–27.

Edwards, W., R.F. Miles Jr., and D. von Winterfeldt, eds. 2007. *Advances in Decision*

Analysis: From Foundations to Applications. Cambridge: Cambridge University Press.

Fabbri, K.P. 1998. A methodology for supporting decision making in integrated coastal zone management. *Ocean & Coastal Management* 39(1–2):51–62.

French, S., and J. Geldermann. 2005. The varied contexts of environmental decision problems and their implications for decision support. *Environmental Science & Policy* 8(4):378–391.

Gieryn, T. F. 1995. Boundaries of science. In *Handbook of Science and Technology Studies*, ed. by S. Jasanoff, T. Pinch, J.C. Petersen, and G.E. Markle, 393–443. Thousand Oaks, CA: Sage Publications.

———. 1999. *Cultural Boundaries of Science: Credibility on the Line.* Chicago: Chicago University Press.

Glasser, R.D. 1995. Linking science more closely to policy-making: Global climate change and the national reorganization of science and technology policy. *Climatic Change* 29(2):131–143.

Global Environmental Assessment Project (GEA). 1997. A critical evaluation of global environmental assessment: The climate experience: A report of the first workshop on global environmental assessment and public policy. Paper read at a workshop convened jointly by the Committee on the Environment of Harvard University, the Center for the Application of Research on the Environment (CARE) of the Institute of Global Environment and Society, Inc., and the International Institute for Applied Systems Analysis, June 1997, at Calverton, MD.

Hall, D., and E. Lobina. 2004. Private and public interests in water and energy. *Natural Resources Forum* 28(4):268–277.

Healey, M.C., and W. Ascher. 1995. Knowledge in the policy process: Incorporating new environmental information in natural resources policy making. *Policy Sciences* 28:1–19.

Healey, M.C., and T.M. Hennessey. 1994. The utilization of scientific information in the management of estuarine ecosystems. *Ocean & Coastal Management* 23(2):167–191.

Herrick, C.N., and J.M. Pendleton. 2000. A decision framework for prediction in environmental policy. In *Prediction: Science, Decision making and the Future of Nature*, ed. by D. Sarewitz, R. Pielke Jr., and R. Byerly Jr., 341–358. Washington, DC: Island Press.

Inter-American Institute for Global Change Research (IAI). 2005. *Linking the Science of Environmental Change to Society and Policy: Lessons from Ten Years of Research in the Americas.* (Group D on Communicating Science). Workshop report from a meeting in Ubatuba, Brazil, 27 November–2 December 2005. São José dos Campos: IAI.

Jacobs, K., G. Garfin, and M. Lenart. 2005. More than just talk: Connecting science and decisionmaking. *Environment* 47(9):6–21.

James, F.C. 1999. Lessons learned from a study of habitat conservation planning. *BioScience* 49(11):871–874.

Jones, S.A., B. Fischhoff, and D. Lach. 1999. Evaluating the science-policy interface for climate change research. *Climatic Change* 43:581–599.

Kann, A., and J.P. Weyant. 2000. Approaches for performing uncertainty analysis in large-scale energy/economic policy models. *Environmental Modeling and Assessment* 5(1):29–46.

Karl, H.A., L.E. Susskind, and K.H. Wallace. 2007. A dialogue, not a diatribe: Effective integration of science and policy through joint fact finding. *Environment: Science and Policy for Sustainable Development* 49(1):20–34.

Kling, G.W., K. Hayhoe, L.B. Johnson et al. 2003. *Confronting Climate Change in the Great Lakes Region: Impacts on Our Communities and Ecosystems.* Cambridge, MA, and Washington, DC: Union of Concerned Scientists and the Ecological Society of America.

Korfmacher, K.S. 1998. Invisible successes, visible failures: Paradoxes of ecosystem management in the Albemarle-Pamlico Estuarine Study. *Coastal Management* 26(3): 191–212.

Lemos, M.C., and B.J. Morehouse. 2005. The co-production of science and policy in integrated climate assessments. *Global Environmental Change* 15(1):57–68.

Lempert, R.J., S.W. Popper, and S.C. Bankes. 2002. Confronting surprise. *Social Science Computer Review* 20(4):420–440.

———. 2003. *Shaping the Next One Hundred Years: New Methods for Quantitative, Long-Term Policy Analysis.* Santa Monica, CA: RAND.

Lindblom, C.E. 1980. *The Policy-Making Process.* 2nd ed. Englewood Cliffs, NJ: Prentice Hall.

Malone, T.F., and G.W. Yohe. 2002. Knowledge partnerships for a sustainable, equitable, and stable society. *Journal of Knowledge Management* 6(4):368–378.

McNie, E.C. 2007. Reconciling the supply of scientific information with user demands: An analysis of the problem and review of the literature. *Environmental Science & Policy* 10:17–38.

Mitchell, R.B., W.C. Clark, D.W. Cash, and N. Dickson, eds. 2006. *Global Environmental Assessments: Information, Institutions, and Influence.* Cambridge, MA: The MIT Press.

Morgan, G., H. Dowlatabadi, M. Henrion et al. 2009. *Best Practice Approaches for Characterizing, Communicating, and Incorporating Scientific Uncertainty in Decisionmaking.* A Report by the U.S. Climate Change Science Program and the Subcommittee on Global Change Research. SAP 5.2. Washington, DC: NOAA.

Morgan, M.G., and M. Henrion. 1990. *Uncertainty: A Guide to Dealing with Uncertainty in Quantitative Risk and Policy Analysis.* New York: Cambridge University Press.

Morgan, M.G., M. Kandlikar, J. Risbey, and H. Dowlatabadi. 1999. Why conventional tools for policy analysis are often inadequate for problems of global change. *Climatic Change* 41:271–281.

Moser, S.C. 1997. Mapping the territory of uncertainty and ignorance: Broadening current assessment and policy approaches to sea-level rise. Dissertation, Graduate School of Geography, Clark University, Worcester, MA.

———. 1998. *Talk Globally, Walk Locally: The Cross-Scale Influence of Global Change Information on Coastal Zone Management in Maine and Hawai'i.* Cambridge, MA: John F. Kennedy School of Government, Harvard University.

———. 2005. Which uncertainties matter for decision-making? Development of an integrative decision-centered screening tool with an application to coastal management in California (Poster). U.S. Climate Change Science Program Workshop on "Climate Science in Support of Decision Making," Arlington, VA, November 14–16.

———. 2006. *Climate Scenarios and Projections: The Known, the Unknown, and the Unknowable As Applied to California.* Synthesis report of a workshop, "Elements of Change," held at the Aspen Global Change Institute, 11–14 March 2004, in Aspen, Colorado. Aspen, CO: AGCI.

———. 2009. Whether our levers are long enough and the fulcrum strong? Exploring the soft underbelly of adaptation decisions and actions. In *Living with Climate Change: Are There Limits to Adaptation?*, ed. by W.N. Adger et al., 313–343. Cambridge: Cambridge University Press.

Moser, S.C., D.W. Cash, and W.C. Clark. 1998. Local response to global change: Strategies for information transfer and decision making for cross-scale environmental risks. Paper read at Local Response to Global Change: Strategies for Information Transfer and Decision Making for Cross-Scale Environmental Risks, 1998, at BCSIA, Harvard University, Cambridge, MA.

Moss, R.H. 2007. Improving information for managing an uncertain future climate. *Global Environmental Change* 17(1):4–7.

Moss, R.H., and S.H. Schneider. 2000. Uncertainties in the IPCC TAR: Recommendations to lead authors for more consistent assessment and reporting. In *Guidance Papers on the Cross Cutting Issues of the Third Assessment Report of the IPCC*, ed. by R. Pachauri, T. Taniguchi, and K. Tanaka, 33–51. Geneva: World Meteorological Organization.

National Council for Science and the Environment. 2000. *Recommendations for Improving the Scientific Basis for Environmental Decisionmaking: A Report from the First National Conference on Science, Policy, and the Environment.* Washington, DC: National Academy of Sciences.

National Research Council (NRC). 1981. *Managing Climatic Resources and Risks.* Report from the Panel on the Effective Use of Climate Information in Decision Making. Washington, DC: National Academy Press.

———. 1989. *Improving Risk Communication.* Washington, DC: National Academy Press.

———. 1996. *Understanding Risk: Informing Decisions in a Democratic Society.* Washington, DC: National Academy Press.

———. 2005. *Decisionmaking for the Environment: Social and Behavioral Science Research Priorities.* Washington, DC: National Academy Press.

———. 2006. *Completing the Forecast: Characterizing and Communicating Uncertainty for Better Decisions Using Weather and Climate Forecasts.* Washington, DC: National Academy Press.

———. 2009. *Informing Decisions in a Changing Climate.* Washington, DC: National Academies Press.

National Research Council, Commission on Geosciences. 1994. *Environmental Science in the Coastal Zone: Issues for Further Research.* Washington, DC: National Academy Press.

National Research Council Roundtable on Science and Technology for Sustainability. 2005. *Knowledge-Action Systems for Seasonal to Interannual Climate Forecasting: Summary of a Workshop.* Washington, DC: National Academies Press.

Nutley, S.M., I. Walter, and H.T.O. Davies. 2007. *Using Evidence: How Research Can Inform Public Services.* Bristol, UK: The Policy Press.

Patt, A. 2007. Assessing model-based and conflict-based uncertainty. *Global Environmental Change* 17(1):37–46.

Patt, A., and S. Dessai. 2005. Communicating uncertainty: Lessons learned and suggestions for climate change assessment. *Comptes Rendus Geosciences* 337(4):425–441

Pielke Jr., R.A. 1997. Policy for science for policy: A commentary on Lambright on ozone depletion and acid rain. *Research Policy* 26:157–168.

Pielke, Jr., R.A. 2001. Room for doubt. *Nature* 410 (8 March 2001): 151.

Pielke Jr., R.A., and R.T. Conant. 2003. Best practices in prediction for decision-making: Lessons from the atmospheric and earth sciences. *Ecology* 84(6):1351–1358.

Pulwarty, R.S. 2003. Climate and water in the West: Science, information and decision-making. *Water Resources Update* 124:4–12.

Pulwarty, R.S., and T.S. Melis. 2001. Climate extremes and adaptive management on the Colorado River: Lessons from the 1997–1998 ENSO event. *Journal of Environmental Management* 63:307–324.

Sarewitz, D., R. Pielke Jr., and R. Byerly Jr. 2000. *Prediction: Science, Decision Making and the Future of Nature.* Washington, DC: Island Press.

Sarewitz, D., and R.A. Pielke. 2007. The neglected heart of science policy: Reconciling supply of and demand for science. *Environmental Science & Policy* 10:5–16.

Steel, B., P. List, D. Lach, and B. Shindler. 2004. The role of scientists in the environmental policy process: A case study from the American West. *Environmental Science & Policy* 7(1):1–13.

Swets, J.A., R.M. Dawes, and J. Monahan. 2000. Better decisions through science. *Scientific American* (October):82–87.

Tribbia, J., and S.C. Moser. 2008. More than information: What coastal managers need to plan for climate change. *Environmental Science & Policy* 11(4):315–328.

Vogel, C., S.C. Moser, R.E. Kasperson, and G.D. Dabelko. 2007. Linking vulnerability, adaptation, and resilience science to practice: Pathways, players, and partnerships. *Global Environmental Change* 17(3–4):349–364.

von Kerkhoff, L. 2005. Integrated research: Concepts of connection in environmental science and policy. *Environmental Science & Policy* 8:452–463.

von Kerkhoff, L., and L. Lebel. 2006. Linking knowledge and action for sustainable development. *Annual Review of Environmental Resources* 31:445–477.

Webster, M. 2003. Communicating climate change uncertainty to policy-makers and the public. *Climatic Change* 61(1–2):1–8.

Wynne, B. 1992. Uncertainty and environmental learning: Reconceiving science and policy in the preventive paradigm. *Global Environmental Change* 6(2):87–101.

Linking Science to Decision Making in the Great Lakes Region

JOEL D. SCHERAGA

HOW DOES ONE PROVIDE TIMELY AND USEFUL SCIENTIFIC INFORMA-tion about climate change to decision makers in the Great Lakes region so they can make more informed decisions? Decision makers and resource managers are becoming increasingly aware that climate change may have important implications for the work they do and the attainment of their goals, and should therefore be an additional consideration in their decision-making processes. They understand the need to anticipate and adapt to a changing climate. Consequently, there is a growing demand for scientific information, data, models, and tools to inform and facilitate decisions.

There has been a rush of activity by the climate-research community to meet this demand for "decision support." Efforts are underway to pro-duce decision-support tools, models, and information that will empower resource managers to account for future changes in climate and make more informed decisions. But these efforts have, to some degree, been distracted by debates over the feasibility of producing scientifically sound projections of future climate and its associated impacts. Some have argued that it is not yet possible to produce projections of climate change at a fine enough geo-graphic resolution to be useful for informing policy decisions. Others have argued that "cascading uncertainties" through end-to-end assessments of the impacts of climate change render the results useless.

These debates are misleading. A wealth of information about climate change and the risks it poses to human health, the environment, and social

---

The views expressed are the author's own and do not represent official EPA policy.

well-being is available. This includes, for example, plausible scenarios of future climate change that have been "downscaled" to a fine enough geographic resolution to be useful for a variety of purposes (Bader et al. 2008). Whether or not this information is useful depends on the specific questions being asked by the decision makers, the types of insights they are seeking, the time frame in which the decisions will be made and the information is needed, and the levels of uncertainty that are acceptable to the decision maker. (Ultimately, it is up to the decision maker to decide "how good is good enough.")

This paper focuses on three key messages: (1) It is now possible to conduct scientifically sound, place-based assessments of the potential impacts of climate change that could inform decision makers in the Great Lakes region; (2) decision support resources other than assessments exist that can be used by resource managers to incorporate considerations of climate change into their day-to-day operations; and (3) it is essential that developers of decision support resources engage the end users to ensure the timeliness and effectiveness of the products.

FEASIBILITY OF CONDUCTING SCIENTIFICALLY SOUND PLACE-BASED ASSESSMENTS

There has been considerable debate in the scientific and policy communities about the feasibility of conducting scientifically sound place-based assessments of the potential impacts of climate change. Much of this skepticism is based on real and perceived limitations of "general circulation models" (GCMs) of the climate system that are used to project future changes in climate (O'Keefe and Kueter 2004; Kerr 1997; U.S. General Accounting Office 1995), as well as the process of downscaling GCM output to regional and local scales with enough resolution to credibly discern potential impacts (Kerr 2000). The argument has been made that GCMs cannot yet be used to make believable predictions of future changes in climate, and therefore cannot be used to conduct impacts assessments. For example, O'Keefe (2004) has stated that "long-term regional assessments simply cannot be done at this time." Still others have argued that large uncertainties that "cascade" through integrated modeling systems that start with GCMs and end with impacts models (Webster et al. 2001; Schneider 2001) render the results of any climate-impacts assessment suspect, meaningless, or useless (Idso and Idso 2001).

These arguments are flawed. The science of climate modeling has matured. Plausible scenarios of future climate change exist. The models now include a greater number of physical processes, project changes at a finer spatial resolution, and successfully simulate a growing set of processes and phenomena (Bader et al. 2008). State-of-the-science models now show many consistent features in their simulations and projections for the future. At the same time, simulations from the different models have not fully converged. This is not surprising, since different modelers approach uncertain model aspects in distinctive ways. No single model is superior to others in all respects. Different models have different strengths and weaknesses. For this reason, analyses of the potential impacts of climate change should rely on an ensemble of climate models and plausible scenarios of climate change.

The "cascading uncertainty" argument is also flawed. As William Easterling notes in another chapter in this volume, climate change is an uncertain science, and "one of the formidable challenges to the efficient exchange of usable knowledge concerning climate change adaptation between the research community and the managers, policy makers, and other stakeholders who would benefit from it is the large inherent uncertainty in that knowledge. But it is not reasonable to hold all scientific knowledge to a standard of indisputable fact in order for it to be considered usable knowledge."[1]

"End-to-end" assessments of the impacts of climate change that start with GCM output and end with impacts models provide information that can inform a variety of resource-management decisions. Whether or not particular modeling uncertainties are even relevant depends upon the question being asked by the decision maker. Also, it is up to the decision maker to decide whether the degree of scientific uncertainty is acceptable. It is certainly the responsibility of the scientist conducting an assessment to ensure that uncertainties are identified and quantified, and to characterize the implications of the uncertainties for the decisions being informed. But once this has been done, it is the responsibility—and prerogative—of the decision maker to decide whether the information produced by a particular assessment is useful. This determination will depend upon a variety of factors, including the risk aversion of the decision maker, the particular question being asked, the urgency with which the information is needed, and the time frame within which the decision must be made.

In addition to end-to-end impacts assessments, there are other types of analyses that can provide useful insights. Different analytic approaches, with different strengths and weaknesses, can be used to answer different questions being asked by decision makers. That is, different analytic approaches can provide different types of information with different levels of uncertainties. The challenge is to choose the right analytic approach for the right question, i.e., an appropriate analytic approach given the particular question being asked by the decision maker.

The projections (scenarios) of future climate provided by GCMs are useful story lines,[2] and the range of story lines allows examination of a range of plausible futures that cannot be examined experimentally (National Assessment Synthesis Team 2000). But analytic approaches other than those that use GCMs exist, such as the use of historic analogs, bounding exercises, and "What if?" scenario analyses. The choice of analytic approach should depend upon the particular questions being asked by the decision maker (i.e., the client for the information produced by the assessment).

Decision makers are often interested in a wide array of information about what climate change may mean for the resources under their purview and the decisions they have to make. For example, a resource manager who is first being introduced to the issue of climate change may simply want to know whether or not the resource he or she manages is sensitive to changes in climate, so as to determine whether climate change warrants further consideration. Simple bounding analyses may be sufficient for providing this information. Other decision makers (e.g., public health officials) may already recognize that certain health outcomes are sensitive to changes in climate (e.g., deaths during heat waves due to heat stress), but may wish to understand the extent to which people are still vulnerable despite the existing public-health infrastructure and mechanisms for responding to climate extremes. Examination of historic analogs could provide this information. In still other cases, a decision maker may already be convinced that climate change is an issue of concern despite the existence of uncertainties, but may want to know whether it is still sensible to wait for the science to improve before making immediate investments in adaptation. He or she may wish to understand whether delaying investments will foreclose options for adapting to climate change in the future and lead to potentially costly and undesirable outcomes. The analysis of "What if?" scenarios may help inform this decision maker.

Given the diversity of information needs by decision makers, the first step in an assessment process should be the identification of the specific issues of concern to the decision maker, and the time frame within which information is needed. Once these issues are understood, the analytic techniques most appropriate for answering the decision maker's questions in a timely fashion can be determined.[3]

CASE STUDY: CLIMATE CHANGE AND COMBINED SEWER OVERFLOW EVENTS

To illustrate the type of timely assessment that can be done to inform decision makers, consider city managers who are already redesigning aging combined sewer systems (CSS) to comply with the EPA's Combined Sewer Overflow (CSO) Control Policy (EPA 1994).[4] Billions of dollars will soon be invested in the rebuilding of CSSs or the development of alternative mechanisms to prevent the overflow of CSSs (e.g., the use of permeable surfaces in surrounding areas to mitigate runoff into the systems).

This is an issue of particular concern to the Great Lakes region. There are combined sewer systems in about 770 communities serving approximately 40 million people across the United States (EPA 2006). Communities in the Great Lakes region contain 182 CSSs, which are generally found in older cities and towns (figure 1).[5] CSSs were among the earliest sewers built in the United States and continued to be built until the middle of the twentieth century. These systems collect and co-treat stormwater and municipal waste water, and are designed to overflow directly to surface waters when their design capacity is exceeded. During intense storms, the capacity of combined sewer systems can be exceeded, resulting in the discharge of untreated stormwater and waste water into receiving streams.

Some CSSs overflow infrequently; others, with every precipitation event. CSO discharges of untreated sewage and stormwater are estimated at 1,260 billion gallons per year nationwide (EPA 2001).

Figure 1. Combined sewer system communities
Source: EPA.

It will be expensive to redesign the CSSs. It is estimated (EPA 2001) that 44.7 billion dollars (in 1996 dollars) in future investments will be needed to control CSOs. Given the magnitude of this investment, city planners want to ensure that the redesigned systems attain their desired performance levels even as the climate changes. Once the systems are rebuilt, they will have long lifetimes—on the order of 100 years—and will be costly to retrofit.

The problem is that climate change is already leading to an increase in the proportion of intense rainfall events (figure 2; U.S. Global Change Research Program 2009). This trend is expected to continue as the earth warms and the hydrologic cycle intensifies, and these intense rainfall events can cause CSSs to overflow. The question facing city planners is whether or not climate change will reduce the likelihood that the redesigned CSSs will satisfy EPA's Control Policies if they are rebuilt without the potential effects of climate change factored into the system designs.

Answering this question is a complex problem because the potential implications of climate change for CSO events in cities across the country are site-specific. Whether or not climate change will lead to an increase in the frequency of CSO events depends upon the particular city and combined sewer system under consideration, and the expected climate at that location.

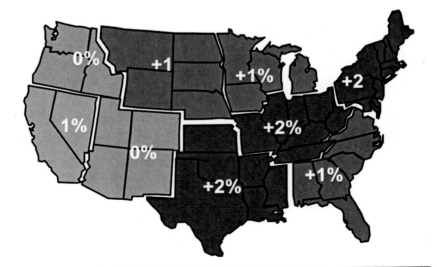

Figure 2. Trends in proportion of annual precipitation of extreme intensity (i.e., more than 2" per day), 1910–1995
Source: Karl and Knight (1998).

Given the high number of CSSs in the Great Lakes region, EPA's Global Change Research Program conducted an assessment of the potential impacts of climate change on CSOs in the Great Lakes region (EPA 2008). The assessment framed the problem in the following way: Suppose each community in the Great Lakes region redesigns their system to achieve an average of four CSO events per year (EPA's so-called "presumption approach" threshold). But also suppose they base new system designs on historical precipitation data, and fail to consider potential changes in future climate. When the climate changes, how might CSO event frequency change, and in how many cases will the four CSO events per-year threshold be exceeded? The purpose of this analysis was to inform city planners about the extent to which future climate change should be a concern as they redesign and rebuild their CSSs.

The first step in the analysis was to determine a "benchmark storm event" for each CSS community; that is, the largest storm event that will need to be captured by the redesigned CSS to meet the four-event-per-year average. This was done using historical precipitation data sets developed for the Vegetation-Ecosystem Modeling and Analysis Project (VEMAP).[6] The second step was to project future storm events. In the VEMAP analysis, two GCMs were used to project future conditions: the Hadley Centre Model and the Canadian Climate Centre Model.[7]

The historical and projected precipitation data were analyzed using both a 1-day and a 4-day moving average. The moving-average approach was used to bracket the effects of short, intense storms, as well as longer storms or multiple precipitation events that might occur in sequence. The historical precipitation data for the 40-year period, 1954 to 1993, were then compared to the projected precipitation data for the future 40-year period, 2060 to 2099, for each CSS community.

The results of this analysis provided insights for the Great Lakes region that were robust across both GCM models. Relative to an assumed four-event-per-year benchmark event, the average annual CSO frequency across the entire Great Lakes region would increase between 13 percent (Canadian model, 1-day averaging period) and 70 percent (Hadley Centre model, 4-day averaging period). In other words, the average number of CSO events per year per CSS community would increase to 4.5 (from the 4-events standard) using the lower bound, and 7.1 (from the 4-events standard) using the upper bound. Across all 182 CSS communities and GCMs, this translates to about 237 events per year above the objectives of EPA's Control Policy. (There is, however, variation among CSS communities—further emphasizing the need to do site-specific studies.)[8]

The results of this "scoping" study provide important (and scientifically sound) insights to city planners:

1. Climate change will affect the future performance of many CSSs in communities throughout the Great Lakes region. City planners need to ascertain the extent to which climate change poses a risk that redesigned systems in their CSS communities will not be effective in meeting the Control Policy's threshold of no more than four CSO events per year.
2. In communities where climate change poses a risk, engineers should not base their calculations of system size (e.g., storage capacity) on current hydrology and historic precipitation data.
3. In those communities where climate change may lead to exceedances above the objectives of EPA's Control Policy, city planners must decide whether to make the additional investments necessary to build in an additional margin of safety to ensure the future effectiveness of the new CSSs. This policy decision will partly depend upon the risk aversion of city planners; that is, the extent to which they are willing to accept the risk that they will incur significant costs to retrofit or refurbish the redesigned CSSs in the future.
4. The risks posed by climate change to CSSs are manageable. It is possible to anticipate the effects of climate change on CSSs and to adapt their new designs to increase the likelihood that they will be effective in the future. Once the city planners make a decision about the level of risk they are willing to incur, engineers can adjust the system designs accordingly to account for climate change.

This example demonstrates how valuable information about the risks posed by climate change can, in fact, be produced in a timely fashion for a decision maker using a scientifically sound approach. Combined sewer systems are being redesigned today, and major investments will soon be made to rebuild the systems. These investments will be made whether or not the scientific community is prepared to inform the city planners about the potential effects of climate change on future system performance. In the EPA study, a screening analysis that appropriately used GCM output provided timely information to city planners making decisions today about the importance of considering climate change as they make major investments in long-lived capital.

BEYOND ASSESSMENTS: A WIDE ARRAY OF DECISION SUPPORT RESOURCES[9]

The term "decision support" reflects the broad purpose of making scientific knowledge about climate change more readily available and more useful to decision makers, in organizations at various levels of government and in the private and nonprofit sectors. It is presumed that by making scientific information about climate variability and change and its potential consequences more timely and useful, management and policy decisions will be better informed, and societal outcomes will be improved.

Despite the broad common goal, there are diverse notions about what decision support and related concepts such as assessment and science application entail and, consequently, about which activities can best achieve the common goal. For example, some think of decision support primarily from the perspective of the producers of the scientific knowledge, and see it as mainly involving translations, other transformations, and communication of available scientific information. Others think of decision support primarily from the perspective of the potential users, and see it mainly as a way of improving the ability of decision makers to take advantage of the information scientific analysis can offer. Still others see decision support as an interaction between science producers and users that can in principle transform both the information that environmental science produces and the ways decision makers use such information (Moser, this volume; NRC 1996).

Given this diversity of views, the notion of "decision support resources" has emerged to capture the wide array of approaches to providing decision support. The U.S. Climate Change Science Program (CCSP) defined "decision support resources" as "the set of analyses and assessments, interdisciplinary research, analytical methods (including scenarios and alternative analysis methodologies), model and data product development, communication, and operational services that provide timely and useful information to address questions confronting policymakers, resource managers and other stakeholders" (U.S. Climate Change Science Program 2003).

The efforts of federal agencies that participate in the CCSP (now known as the U.S. Global Change Research Program) reflect the diversity of approaches to providing decision support. This diversity may reflect their different missions and the different constituencies, decision makers, and types of problems they face. Agencies may devote decision support resources to (1) transforming technical presentations of available scientific information into forms they believe will be more easily understood by potential users; (2) analyzing the sensitivity or adaptive capacity of particular regions or sectors to climatic events so as to direct research efforts differentially to areas where the potential is greatest for knowledge that can lead to improved adaptive management; (3) establishing dialogues between the producers and users of research through which the producers can adjust their research priorities to better meet users' needs; (4) developing tools that practical decision makers can use to interpret available scientific information or make sense of scientific uncertainty in the context of their decisions; and (5) analyzing decisions in particular sectors and the associated information needs in order to develop scientific information targeted to meeting the needs.

The above list illustrates the variety of possible activities that decision support systems might include. It also suggests the need for the research community to develop a clearer conceptual and operational framework for designing and evaluating decision support activities. Such a framework could help public and private institutions interested in developing decision support resources think systematically about types of decisions and decision makers, decision contexts (factors associated with decisions faced by particular kinds of decision makers in particular spatial and temporal frameworks), kinds of decision support that might be provided, and ways to map decision contexts against decision support needs. A framework that supported systematic thinking of this kind could help institutions identify and systematically consider the decision support objectives for specific contexts. It would enable institutions to consider decision support tools in light of these objectives, judge in a more systematic way which objectives are most critical to pursue in supporting decision makers in a particular sector, consider how the various activities of a decision support program can be organized to contribute to better decisions, and develop defensible methods for evaluating their decision support efforts.[10]

CASE STUDY: EPA'S BASINS CLIMATE ASSESSMENT TOOL FOR WATER RESOURCE MANAGERS

Climate change during the next century is likely to result in warmer temperatures, changes in the amount of precipitation, and increases in the intensity of precipitation throughout much of the United States. Water resources and aquatic ecosystems are highly vulnerable to these changes, with possible effects including increased occurrence of floods and droughts, water-quality degradation, channel instability and habitat loss, and impacts on aquatic biota. Thus climate change presents a significant risk of disruption to many water systems. A better understanding of climate change impacts would improve the ability of water managers to meet future supply needs, comply with water-quality regulations, and protect aquatic ecosystems.

In response to this need, EPA's Global Change Research Program developed a Climate Assessment Tool that enables water resource managers and other stakeholders to incorporate considerations of climate change—along with the myriad other stressors with which they are concerned—into their decision-making processes (EPA 2009). The tool facilitates the evaluation of potential adaptation strategies to increase the resilience of water systems to change. It is adaptable to specific watersheds, which is important given the unique nature of most watersheds and water systems. And it can account for stressors such as land-use/land-cover change and point-source loading together with climate. A key advantage of this tool is the capability it provides for the user to define a wide range of hydrologic and water quality endpoints (e.g., 100-year flood events; annual nutrient load to a river), and to systematically assess the sensitivity of the user-defined endpoints to potential changes in climate.[11]

The Climate Assessment Tool (CAT) is embedded in EPA's BASINS modeling system. BASINS is a well-documented, widely distributed tool for decision support in watershed management. As such, it offered a unique platform upon which to develop a Climate Assessment Tool that would be useful to stakeholders concerned with climate change.

BASINS/CAT can be used to conduct standard assessments ("What would be the effect of climate change on X?") and sensitivity analyses ("What would we have to believe is true for X to happen?"). Sample applications of BASINS/CAT to conduct standard assessments include

1. an assessment of how increases in precipitation of 10, 20, and 30 percent, etc., over historical conditions will influence non-point pollution loading to a stream;
2. an assessment of the flooding that would be caused by a historical extreme weather event given recent increases in urban development within the watershed; and
3. an assessment of the future effectiveness of a proposed TMDL implementation plan under a projected climate change scenario.

Sample applications of BASINS/CAT to conduct sensitivity assessments include (1) determination of the change in annual precipitation or timing of runoff that would need to occur to require revision of a reservoir operation

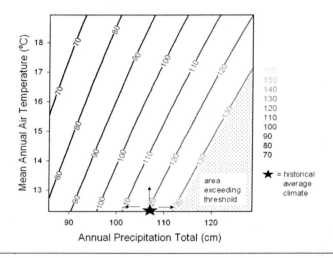

Figure 3. BASINS/CAT sensitivity analysis of annual nitrogen loading to changes in temperature and precipitation
Source: Personal communication with Dr. Thomas Johnson (EPA).

plan; (2) determination of the change in precipitation event intensity that would need to occur to require revision of a TMDL implementation plan; and (3) determination of the change in mean annual air temperature that would increase stream temperatures to the point where trout cannot survive and reproduce.

ENGAGEMENT OF END USERS FOR EFFECTIVE DECISION SUPPORT

I have argued that an array of decision support resources can be developed that can help make scientific information about climate change and its potential consequences more timely and useful. I return now to the importance of developing decision support resources in partnership with the end users to ensure they are timely and useful (i.e., they address the particular needs and questions being asked by the decision makers).[12]

Altalo (2005) has emphasized the critical importance of understanding the end users and how they do business. This includes understanding who are the users, how they are institutionally organized (which may limit the types of decision support they can use or are willing to use), why they need the resources, and how they will use the resources. The degree to which a decision support activity is successful depends upon the assumptions made about the nature of decision-making processes (Pyke and Pulwarty 2006). An adequate level of understanding of the decision-making processes can only be obtained through a dialogue between producers of decision support and the end users.

An ongoing dialogue is essential. For decision support to be *informative*, the providers of the support must know the particular issues of concern to the stakeholders (i.e., those parties with the ultimate interest in the consequences of a problem or its solution), the questions they would like answered, and the tools they need developed. This requires that stakeholders be engaged from the outset of the assessment process, and increases the likelihood that the decision support resources will be effective (National Research Council 2008; Scheraga and Furlow 2001).[13]

In some cases, it might be impossible to know how to target the development of decision support resources unless input is obtained from the end users. Consider, for example, a decision support activity intended to develop indicators of ecosystem health that can be monitored as the climate changes. Whether or not a forest ecosystem can be considered "healthy" is a function of end uses. A forest that would be considered healthy by campers and hikers might be considered unhealthy by a commercial timber company. Consequently, in the absence of stakeholder input, it is impossible for the decision support community to know which indicators of ecosystem health are the most appropriate to develop.

Stakeholders should also be involved in the decision support process on an ongoing basis (Scheraga and Furlow 2001). At a minimum, including the stakeholders in the process makes the decision support resources that are produced more transparent and credible. Also, in many cases, the stakeholder community can offer data, analytic capabilities, insights, and understanding of relevant problems that can contribute to the decision support process. For example, assessors working on the Great Lakes Regional Assessment that was conducted as part of the U.S. National Assessment (Sousounis et al. 1998) discovered that by working with representatives of the ski industry, they were able to gather valuable information and insights on the impacts of reduced snowfall on winter sports. Winkler (this volume) provides an update on the ongoing work that was initiated then. Moser (this volume) provides a detailed description of a process that has been used successfully in California coastal planning.

For a decision support activity to be *timely*, providers of decision support must understand how the information will be used by the relevant stakeholders, and the time frame within which the information is needed (see Moser, this volume). Even with stakeholder involvement, scientists often are hesitant to make definitive statements that might be used by policymakers, because scientific uncertainties still exist; the science is not yet "perfect." Yet, policymakers often have to make decisions under uncertainty, whether or not scientists are prepared to inform those decisions. Providers of decision support must strive to answer decision makers' questions to the extent possible given uncertain science, characterize the uncertainties, and explain their implications for different policy or resource management decisions.

Informed decisions are better than uninformed decisions.

NOTES

The author is grateful to Tom Johnson for his assistance with the material included under Case Study: EPA's BASINS Climate Assessment Tool for Water Resource Managers.

1. In another chapter in this volume, Marx and Weber note that humans have a great need for predictability. The existence of uncertainty can become a barrier to individuals taking action. The authors provide suggestions on how one might use insights from behavioral decision research in constructive ways to effectively communicate the risks posed by climate change, and design effective decision environments that can lead to policy interventions.

2. GCM output is not a *prediction* of future climate.

3. It's interesting to note that even large corporations in the private sector are beginning to adopt a more customer-oriented approach to R&D programs. Scientists working for large corporations are increasingly designing R&D programs by first listening to customers, identifying their priority needs, and then fulfilling those needs in a timely manner. At Dow Chemical Co., scientists now work with customers to identify and prioritize a "wish list of products or desired technical characteristics" (Lavelle 2004). They then return to the lab to conduct the necessary research and development activities that lead to the product requested by their customers (thus building in guaranteed demand).

4. The EPA Control Policy established a consistent national approach for controlling discharges from CSOs to the nation's waters through the National Pollutant Discharge Elimination System (NPDES) permit program.

5. The number of CSS communities by state are: Indiana–24, Illinois–34, Michigan–46, Minnesota–3, New York–23, Ohio–47, Pennsylvania–3, and Wisconsin–2.

6. VEMAP documentation can be found at http://www.cgd.ucar.edu/vemap/index.html.

7. Both GCMs provide projections on a grid with intervals of 1 degree latitude and longitude.

8. The results for variations among CSS communities are presented in EPA 2008.

9. The author is grateful to Paul Stern for valuable input on this section of the paper.

10. To address this issue, EPA and NOAA cosponsored the NRC study *Informing Decisions in a Changing Climate* (2009). The study (1) elaborated a framework for considering climate-related decision support objectives and activities, (2) assessed the strengths and limitations of various strategies, activities, and tools, and (3) recommended strategies that CCSP agencies might use for organizing decision support activities.

11. BASINS/CAT was officially released in May 2007 as part of the EPA BASINS system (version 4). It is available free of charge on the Internet.

12. This section complements Susanne Moser's discussion elsewhere in this volume about important criteria that need to be met in order for the science/decision-making interaction to work effectively.

13. There are certainly cases in which resources that were developed without any initial engagement of end users proved timely and useful to some group of decision makers. But the attitude that "If we build it, they will come" is risky and does not offer a systematic approach to providing effective decision support.

REFERENCES

Altalo, Mary. 2005. "Don't ask me what I want, ask me what I do": The key to valid requirements documentation. PowerPoint presentation at CCSP Workshop on Climate Science in Support of Decision Making, November 14–16, Arlington, VA.

Bader, D.C., C. Covey, W.J. Gutowski et al. 2008. *Climate Models: An Assessment of Strengths and Limitations*. Synthesis and Assessment Product 3.1. Washington, DC: U.S. Department of Energy.

Idso, C.D., and K.E. Idso. 2001. The unstable sands of climatic uncertainty. Editorial commentary. *Center for the Study of Carbon Dioxide and Global Change* 4(23)(6 June).

Karl, T.R., and R.W. Knight. 1998. Secular trends of precipitation amount, frequency, and intensity in the United States. *Bulletin of the American Meteorological Society* 79(2):231–241.

Kerr, R.A. 1997. Climate change: Greenhouse forecasting still cloudy. *Science* 276(5315):1040–1042.

Lavelle, L. 2004. Inventing to order: Dow cuts risk by finding out first what its customers need. *Business Week*, July 5, 84–85.

National Assessment Synthesis Team. 2000. *Climate Change Impacts on the United States: The Potential Consequences of Climate Variability and Change.* Washington, DC: U.S. Global Change Research Program.

National Research Council. 1996. *Understanding Risk: Informing Decisions in a Democratic Society.* Edited by P.C. Stern and H.V. Fineberg. Committee on Risk Characterization. Washington, DC: National Research Council.

National Research Council. 2008. *Public Participation in Environmental Assessment and Decision Making.* Edited by T. Dietz and P.C. Stern. Panel on Public Participation in Environmental Assessment and Decision Making. Committee on the Human Dimensions of Global Change. Washington, DC: The National Academies Press.

National Research Council. 2009. *Informing Decisions in a Changing Climate.* Committee on the Human Dimensions of Global Change. Washington, DC: The National Academies Press.

O'Keefe, W. 2004. The challenge of making science policy relevant. Remarks presented at the EPA Science Forum, June 3.

O'Keefe, W., and J. Kueter. 2004. *Climate Models: A Primer.* Washington, DC: The George C. Marshall Institute.

Pyke, C.R., and R.S. Pulwarty. 2006. Elements of effective decision support for water resource management under a changing climate. *Water Resources Impact* 8(5)(September):8–10.

Scheraga, J.D., and J. Furlow. 2001. From assessment to policy: Lessons learned from the U.S. National Assessment. *Human and Ecological Risk Assessment* 7(5):1227–1246.

Schneider, S.H. 2001. What is "dangerous" climate change? *Nature* 411(May 3):17–19.

Sousounis, P., G. Albercook, J. Andersen et al. 1998. *Climate Change in the Upper Great Lakes Region: A Workshop Report.* Edited by P. Sousounis and G. Albercook. Ann Arbor: University of Michigan.

Sussman, F., and R. Freed. 2006. *Climate Sensitive Decision Making: Towards a Framework for Identifying Good Candidates for Decision Support*. Draft report to the U.S. EPA. Washington, DC: ICF Consulting, March 16.

U.S. Climate Change Science Program. 2003. *Strategic Plan for the U.S. Climate Change Science Program*. A report by the Climate Change Science Program and the Subcommittee on Global Change Research. July.

U.S. Environmental Protection Agency 1994. Combined sewer overflow (CSO) control policy; notice. *Federal Register* 59(75)(Tuesday, April 19):18688–18698.

U.S. Environmental Protection Agency. 2001. *Implementation and Enforcement of the Combined Sewer Overflow Control Policy*. Report to Congress, Office of Water, EPA 833-R-01–003. Washington, DC. December.

U.S. Environmental Protection Agency 2006. *2006–2011 EPA Strategic Plan— Charting Our Course*. September 30, p. 42.

U.S. Environmental Protection Agency. 2008. *A Screening Assessment of the Potential Impacts of Climate Change on Combined Sewer Overflow (CSO) Mitigation in the Great Lakes and New England Regions* (Final Report). EPA/600/R-07/033F.

U.S. Environmental Protection Agency. 2009. *BASINS 4.0 Climate Assessment Tool (CAT): Supporting Documentation and User's Manual* (Final Report). Washington, DC, EPA/600/R-08/088F.

U.S. General Accounting Office. 1995. *Global Warming: Limitations of General Circulation Models and Costs of Modeling Efforts*. GAO/RCED-95–164. Washington, DC. July.

U.S. Global Change Research Program. 2009. *Global Climate Change Impacts in the United States*. Edited by T.R. Karl, J.M. Melillo, and T.C. Peterson. Cambridge, UK: Cambridge University Press.

Webster, M.D., C.E. Forest, J.M. Reilly et al. 2001. *Uncertainty Analysis of Global Climate Change Projections*. MIT Joint Program on the Science and Policy of Global Change. Report No. 73. March (with revisions July 2001).

The Development and Communication of an Ensemble of Local-Scale Climate Scenarios

An Example from the Pileus Project

JULIE A. WINKLER, JEANNE M. BISANZ,
GALINA S. GUENTCHEV, KRERK
PIROMSOPA, JENNI VAN RAVENSWAY,
HARYONO PRAWIRANATA, RYAN S. TORRE,
HAI KYUNG MIN, AND JOHNATHAN CLARK

A RAPIDLY EXPANDING BODY OF LITERATURE FOCUSES ON THE POTEN-tial impacts of future climate change on natural and human systems at spatial scales ranging from global to local. This literature, referred to as assessments of climate change impact, adaptation, and vulnerability (Carter et al. 2007), or more casually as climate change assessments, often has one or more climate scenarios as a starting point. A "scenario" simply refers to an internally consistent and plausible future state (Carter et al. 1996). In climatology, this term is often used interchangeably with "projection," but is carefully distinguished from "prediction" or "forecast." This nuanced distinction is meant to convey that a future climate scenario involves numerous assumptions concerning, among other factors, "future socioeconomic and technological developments that may or may not be realized," and that climate scenarios are "subject to substantial uncertainty" (Mearns et al. 2001, 795).

As summarized by Scheraga (this volume), a number of methods are used to construct future climate scenarios. One approach is to employ historical and/or spatial analogs (e.g., a precipitation reduction similar to a historical unusually dry period, or temperatures observed at a location with a currently warmer climate). Simple "What if?" scenarios (e.g., a 10 percent increase in precipitation or a 2°C increase in mean annual temperature) have also been used in assessment studies. More commonly, climate scenarios are derived from complex, three-dimensional models referred to as general circulation models or, alternatively, global climate models. These models, abbreviated as GCMs, were developed to study climate processes, natural climate variability, and the climatic response to anthropogenic forcing, and have relatively coarse spatial scales (CCSP 2008). Consequently, GCM simulations often need to be downscaled to a finer scale, in some cases to the local scale. Broadly, downscaling procedures fall within two categories, dynamic downscaling and empirical downscaling (Christensen et al. 2007). Dynamic downscaling utilizes projections from GCMs as initial and boundary conditions for a regional climate model with a finer-grid spacing (typically 10 to 50 kilometers) compared to GCMs (typically 150 kilometers or greater). In contrast, empirical downscaling uses statistical relationships derived between larger-scale atmospheric parameters and local or regional climate variables.

The considerable uncertainty surrounding projections of future climate arises from a number of sources including an incomplete understanding of climate processes, future greenhouse gas emissions, and responses of the earth's carbon cycle to changes in greenhouse gas concentrations (Ahmad et al. 2001). As pointed out by several authors of chapters in this volume (i.e., Easterling et al.; Marx and Weber; Moser; and Scheraga), uncertainty is an inevitable part of any decision-making process, and decision makers require information regarding the degree of uncertainty that exists in order to make informed choices and develop alternative strategies. This is certainly the case for climate scenarios, although estimating the degree of certainty is challenging. One approach is to develop an ensemble of climate scenarios, following Jones (2000), who argued that if more than one estimate of a parameter is available (e.g., values from four models as shown for the example in figure 1), the range of values is the "well-calibrated" range of uncertainty. Jones further advocated that scenario developers can define an expanded or "judged" range of uncertainty based on their knowledge of the limitations and biases of the model estimates, and that by increasing the number of estimates and carefully varying the assumptions and procedures for obtaining the estimates, the well-calibrated and judged ranges of uncertainty will better approach the full range of uncertainty.

Other considerations when constructing climate scenarios include the type of climate variables to include and their temporal resolution.

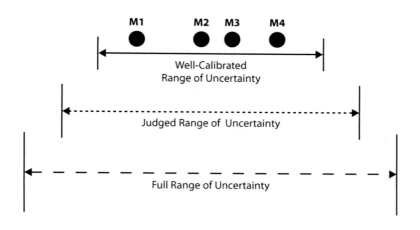

Figure 1. Schematic depiction of the relationship between "well-calibrated" scenarios, the wider range of "judged" uncertainty that might be elicited through decision analytic techniques, and the "full" range of uncertainty, which is drawn wider to represent overconfidence in human judgments. M1 to M4 represent scenarios produced by four models (e.g., globally averaged temperature increases from an equilibrium response to doubled CO_2 concentrations). This lies within a full range of uncertainty that is not fully identified, much less directly quantified by existing theoretical or empirical evidence (modified from Jones, 2000).
Source: Adapted from Ahmad et al. (2001).

Temperature and precipitation are the two most frequently used variables in climate change assessments. For some assessments, annual and/or monthly means are sufficient, whereas for others, daily or even subdaily temperature observations and precipitation totals are needed. In other situations, a more extensive suite of climate variables such as radiation fluxes, humidity, and wind direction and speed may be required. For almost all assessments, the climate scenarios need to be translated into parameters useful to specific stakeholder groups. For example, scenarios of future cooling degree days may be more useful to stakeholders in the energy sector compared to the maximum and minimum temperature scenarios from which the cooling degree days were calculated. Thus, to effectively and efficiently meet the needs of an assessment, the climatologists responsible for the scenario construction must obtain throughout the assessment process detailed input and feedback from relevant stakeholder groups. In this context, stakeholders include other scientists participating in the assessment who will be utilizing the scenarios in their analyses, in addition to the decision makers to whom the assessment is directed and/or has relevance.

Because of the substantial cost and effort involved in constructing a relatively large climate-scenario ensemble, it is desirable to consider a broad spectrum of potential users, beyond those directly affiliated with a particular assessment, when designing the scenarios. However, most previously developed climate scenarios have been used only in the original assessment for which they were constructed. In part, this is the result of not considering other potential users in the initial scenario design, but another impediment is that for climate scenarios to be useful, relevant, and accessible (following the criteria laid out by Moser, this volume), they must be made readily available to scientists and decision makers and accompanied by appropriate and sufficient documentation.

The upshot is that the development and communication of an ensemble of climate scenarios relevant to a broad spectrum of stakeholders is a challenging undertaking. In this chapter, we describe the rationale and philosophy behind the development and communication of climate scenarios for the Pileus Project, a climate change assessment concerned with the impact of future climate on specialized agriculture and tourism in the Great Lakes region. The chapter is intended to provide an example of one approach to this critical component of a climate change assessment, and to hopefully initiate a conversation among scientists and stakeholders on better utilization of the considerable effort and resources expended in climate-scenario development.

THE PILEUS PROJECT: A SHORT INTRODUCTION

The Pileus Project was an outgrowth of the Great Lakes Regional Assessment conducted as part of the U.S. National Assessment. Like the earlier assessment, the Pileus Project was funded by the U.S. Environmental Protection Agency. The project name was taken from the *pileus* cloud, which can appear as a cap above a cumulonimbus cloud, and was chosen to symbolize the "capstone" or overarching nature of the project. The primary goals of the Pileus Project were to (1) identify, with stakeholder assistance, the influence of climate on Michigan's agriculture and tourism industries, (2) create empirical and physically based models to quantify the impacts of past and projected future climate variability and change, (3) develop decision-support tools for weather and climate-related risk management, and (4) build strong stakeholder-researcher partnerships. Because agriculture and tourism in Michigan are broad, multifaceted industries, the project researchers elected to focus on two aspects of agriculture (the tart cherry industry and corn and wheat grain quality) and three aspects of tourism (downhill skiing, golf, and tourism traffic). The choice of these "subindustries" was influenced by their sensitivity to weather and climate, and the importance of the subindustries to Michigan's economy. Project participants included scientists specializing in agricultural economics, climatology, computer science, geography, horticulture, and tourism, as well as stakeholder groups involved in multiple aspects of the two industries, such as individual farmers or resort owners, processors and distributors, and leaders of industry advocacy groups.

The project employed an "end-to-end" strategy, whereby several models are linked in a sequential manner. Climate observations (for a historical period) and climate scenarios (for future periods) were the first link of the sequential modeling system and were fed into an industry-specific simulation model. Depending on the subindustry, the output of the simulation model next served as input to a series of economic and decision-making models.

CLIMATE SCENARIO REQUIREMENTS

At the onset of the project, the team tasked with constructing the climate scenarios worked closely with other project participants to identify the type of climate variables needed, the required spatial and temporal resolution, the future periods of most interest, and methods for developing the scenarios. The following were considered to be either essential or sufficient:

- The spatial scale of the scenarios should be local (rather than regional) to capture important fine-scale influences on climate, such as the impact of the Great Lakes on locations near the lakeshores.
- Representation of the spatial variations in industry output and the potential impact of future climate demanded that scenarios be constructed for multiple locations in the agricultural and tourism regions.
- A daily temporal resolution was necessary for the scenarios to serve as input to crop phenological and yield models.
- Scenarios of the basic climatological variables (maximum temperature, minimum temperature, wet/dry day occurrence, and liquid-equivalent precipitation) were sufficient for most analyses.

- Additional parameters relevant to a particular activity or industry (e.g., thresholds, spells) should be derived from the daily temperature and precipitation series.
- Because of the sensitivity of the agricultural and tourism industries to unusual or extreme events (e.g., late frosts), the scenarios must incorporate potential changes in climate variability.
- Temporal autocorrelation of daily temperature and precipitation should be reproduced reasonably well in the climate scenarios.
- The scenarios should be based on GCM simulations in order to take advantage of the strengths of these models.
- Three 20-year future time slices were of greatest interest to stakeholders:
  - Early century (2010 to 2029)
  - Mid century (2040 to 2059)
  - Late century (2080 to 2099)
- The scenarios should have wider applicability beyond agriculture and tourism.

CONSTRUCTION OF THE SCENARIO ENSEMBLE

As noted above, a number of different methods exist for constructing local climate scenarios. The strengths and weaknesses of the different approaches were carefully considered when selecting the methods for constructing the Pileus Project scenarios. Although used in numerous previous assessments, adjusting long-term daily observations of temperature and/or precipitation by a GCM-predicted change in the annual, seasonal, or monthly means was judged not suitable, as this approach does not allow for possible changes in the variability of the climate parameter. Another approach is to use stochastic methods, popularly called "weather generators," to derive daily time series. One concern of stochastic approaches is that the simulated variability of the synthetic series is usually smaller than the observed variability (Qian et al. 2008). An advantage of dynamic downscaling is that the scenarios are produced by physically based models. On the other hand, regional climate models are time-consuming to run, and consequently it is difficult to obtain a scenario ensemble that can be used to estimate uncertainty. In fact, many, if not most, previous applications of dynamic downscaling have employed only one GCM to derive a regional model (e.g., White et al. 2006), and then usually only for short time slices. A further limitation of dynamic downscaling is that additional downscaling procedures are often needed to obtain local, rather than regional, scenarios.

Given the constraints of the other downscaling methods, empirical downscaling was selected to create the climate scenarios for the Pileus Project. Empirical transfer functions that statistically relate large-scale averages or patterns of one or more predictor variables to local values of a surface climate variable are an outgrowth of statistical forecasting approaches used in real-time weather prediction (e.g., Antolik 2000). The primary advantages of using an empirical downscaling approach are that (1) the scenarios are local in scale, (2) they incorporate variability changes in the GCM simulations of the predictor variables, and (3) the scenario development, although demanding, is less time-consuming than dynamic downscaling, allowing for the development of a scenario ensemble. A limiting assumption of empirical downscaling that needs to be considered when interpreting empirically derived climate scenarios is that the transfer functions are invariant with time.

For the Pileus Project, the statistical transfer functions were derived using free-atmosphere rather than surface variables as predictors. The rationale for this choice is that free atmosphere variables (i.e., variables that are not, or only modestly, influenced by friction near the earth's surface) are generally assumed to be better simulated by GCMs compared to surface variables that are influenced by complex, and often poorly simulated, fluxes between the earth's surface and lower atmosphere (Palutikof et al. 1997). Temperature and precipitation scenarios were constructed separately.

Development of the local temperature scenarios followed the general procedure outlined in figure 2. First, historical maximum and minimum temperature series for 15 locations across Michigan were obtained and inspected for inhomogeneities. Multiple regression was used to relate the daily maximum (or minimum) temperature series (i.e., the predictand) for each station to predictor variables calculated from sea-level pressure and 500 hectopascals (hPa) geopotential height obtained from the reanalysis fields developed jointly by the National Centers for Environmental Prediction (NCEP) and the National Center for Atmospheric Research (NCAR) (Kalnay et al. 1996). The transfer functions were developed using observed values of the predictors and predictands for 1970 to 1989. The functions were then validated using observed values of the predictor variables and predictands for two validation periods: 1960 to 1969, and 1990 to 1999. Next, GCM-simulated values of the predictor variables for 1990 to 1999 (referred to as the "control" period) were interpolated to the station location and input into the transfer functions. The resulting control-period local daily temperature series were validated against observed maximum and minimum temperature for the same period (1990 to 1999). Finally, the transfer functions were applied to GCM-simulated future values of the predictor variables for the period out to 2099, and the future series were constructed. For precipitation, transfer functions were first developed using data-mining classification techniques to project whether a day was wet or dry. Regression methods were then used to simulate the precipitation amount on wet days, similar to the procedures used to develop the temperature transfer functions.

The ensemble of scenarios for each climate variable and location is based on simulations from four GCMs, each driven with two different greenhouse-gas emission scenarios. In addition, multiple variants of the downscaling methods were employed. The four GCMs used to create the scenarios were the CGCM2 model of the Canadian Climate Center, the HadCM3 model developed by the Hadley Center in Great Britain, the ECHAM4 model developed by the Max Planck Institute in Germany, and the CSM1.x model developed by NCAR in the United States. These models

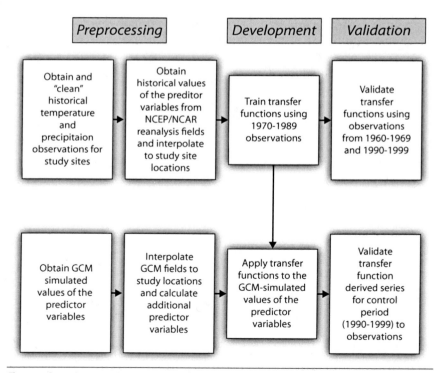

Figure 2. Procedures used to develop the future climate scenarios

were chosen because they have been widely used in climate impact studies, and the modeling groups have made time series of 500 hPa, sea-level pressure, and other free atmosphere variables at a daily time step available to impact researchers. All are Intergovernmental Panel on Climate Change (IPCC) Third Assessment era models; model simulations prepared for the IPCC Fourth Assessment were not yet available to the user community at the start of the Pileus Project.

The IPCC has developed a suite of scenarios on how greenhouse gas emissions may change in the future, which are widely known as the SRES scenarios (Nakicenovic et al. 2000). Only two of the SRES scenarios (A2 and B2) were used in the development of the future-climate scenario ensemble for the Pileus Project. The reason for this is that few modeling groups made available GCM simulations at daily time steps for the full range of SRES scenarios. The A2 scenario describes a heterogeneous world with relatively slow technological change and modest international cooperation. Flows of trade, people, and capital between regions are small compared to other SRES scenarios; fertility rates decline slowly, leading to a large global population at the end of the twenty-first century; and greenhouse gas concentrations increase throughout the twenty-first century, reaching approximately 800 parts per million by 2099 (Nakicenovic et al. 2000). The B2 scenario projects a world that is increasingly concerned with environmental sustainability and emphasizes local solutions to economic, social, and environmental problems. Population growth is substantially smaller for the B2 scenario compared to the A2 scenario, and greenhouse gas emissions stabilize in the latter half of the twenty-first century at around 600 parts per million (Nakicenovic et al. 2000).

These choices allowed for several sources of fundamental and structural uncertainty to be captured in the scenario ensemble. Fundamental uncertainty refers to surprises or complex, chaotic conditions that cannot be modeled (Easterling et al., this volume). Comparison of the local climate scenarios constructed using the two different greenhouse-gas emission estimates allows the fundamental uncertainty from uncertain future greenhouse gas emissions, and underlying political, socioeconomic, and technological futures, to be partially evaluated. Structural uncertainty, on the other hand, results from differences in model structure due to imperfect understanding of a system. The use of multiple GCMs permits an evaluation of the uncertainty introduced by the limited ability to model the climate system.

Differences in downscaling methods introduce another source of structural uncertainty. Earlier work by Pileus Project researchers (Winkler et al. 1997) showed that quasi-subjective decisions made by the climate analyst when designing empirical transfer functions can have an influence on the resulting scenarios. Three "user decisions" were considered when developing the scenario ensemble: (1) the decision to adjust for deviations of the GCM control simulations of the predictor variables from observations, (2) the definition of the seasons for which separate specification equations are derived, and (3) the decision to remove the annual cycle from the predictor variables and predictands. The possible variants are summarized in figure 3.

Notation	Decision #1: Adjust for errors in the GCM control simulations of the predictor variables	Decision #2: Length of the specification period	Decision #3: Annual cycle removed from predictor variables and predictands
PN	No adjustment for errors in the control simulations.	A single transfer function is developed for the entire year.	Annual cycle not removed.
PS	As for PN.	Separate transfer functions developed by season (winter, spring, summer, and fall).	Annual cycle not removed.
ZN	Errors in GCM-simulated predictor variables adjusted by imposing same mean and standard deviation on observed and control GCM series.	As for PN.	Annual cycle not removed.
ZS	As for ZN.	As for PS.	Annual cycle not removed.
PN_D	As for PN.	As for PN.	Annual cycle removed.
PS_D	As for PS.	As for PS.	Annual cycle removed.
ZN_D	As for ZN.	As for ZN.	Annual cycle removed.
ZS_D	As for ZS.	As for ZS.	Annual cycle removed.

Figure 3. Empirical downscaling variants

The end result of the scenario development was more than 60 scenarios for each of four climate variables—daily maximum temperature, daily minimum temperature, occurrence of a wet or dry day, and daily liquid-equivalent precipitation amount—at 15 locations in the Great Lakes region. This climate-scenario ensemble is considerably larger than any used previously, and provided the participants of the Pileus Project with an unprecedented opportunity to estimate the uncertainty surrounding future climate projections.

DELIVERING THE SCENARIOS: A WEB-BASED TOOL

An important goal at the onset of the Pileus Project was to distribute the climate scenarios to stakeholders and other interested parties via a web-based user tool. Although invaluable for many aspects of the Pileus Project, the large size of the scenario ensemble posed a challenge when developing software and designing graphics to communicate the scenarios. Also, because potential users were unlikely to be familiar with much of the terminology and the different aspects of the scenario development, another challenge was to provide sufficient background information in an easy-to-understand and visually appealing format so that stakeholders could use the scenarios in an informed, thoughtful manner in their applications and decision making.

After considerable discussion with the project stakeholders, the user tool focused on presenting projected changes in the median value or frequency of a set (165 total) of climate parameters derived from the daily maximum and minimum temperature, wet/dry day, and liquid-precipitation scenarios. The parameters were categorized by whether they are most relevant to agriculture or to tourism (figure 4). All quantities were expressed in English units, as these are the units most often used by the stakeholder communities. Some examples of parameters included in the tool are the date of last spring freeze, the length of the frost-free season, the number of cooling degree days, the number of hot (95°F and hotter) days, growing degree-day accumulation for different base temperatures, and the frequency of two or more wet days in a row.

After considerable experimentation, the Future Scenarios Tool was designed to provide five major graphical displays (figure 5):

- *Reference Climate.* This display shows observed (1981 to 2000) values for a particular parameter and provides an indication of the year-to-year variability

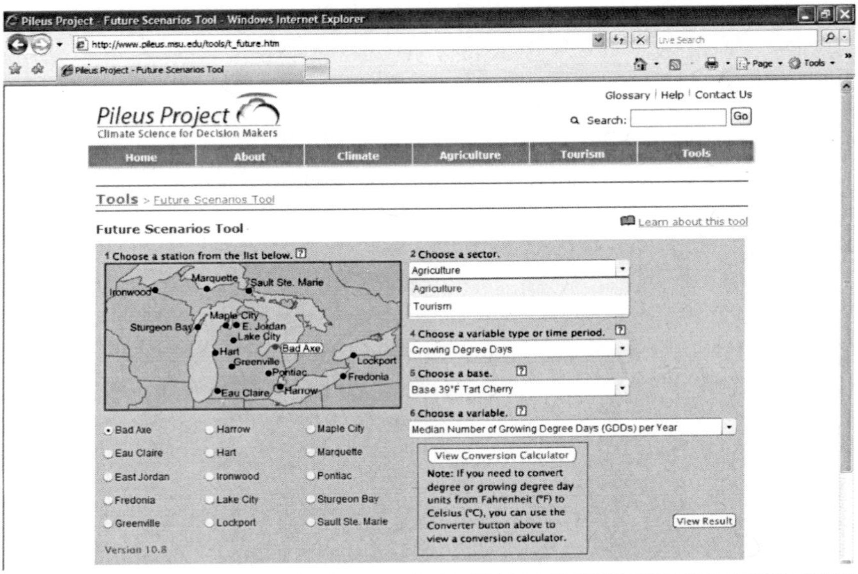

Figure 4. Front page of the Future Scenarios Tool. Users must first choose a location and then either the agriculture or tourism sector.

in the parameter. The user can also click on a button in the lower-right portion of the screen to view the median value of the parameter for the 20-year reference period.

- *Early vs. Mid Century.* When a user chooses this option, a histogram of the projected changes in the median value (or frequency) of a parameter for an early-century period (defined as 2010 to 2029) and a histogram for a mid-century period (defined as 2040 to 2059) are displayed. The spread of the bars for the histograms is an indication of the quantifiable uncertainty range for the two time periods.

- *Mid vs. Late Century.* A similar display is available for mid- and late-century projected changes, with one histogram of the projected changes in the median value (or frequency) of a parameter for a mid-century period (defined as 2040 to 2059) and another of the projected changes for the late-century period (defined as 2080 to 2099). Again, the spread of the histograms indicates the quantifiable uncertainty range for the two time periods.

- *A2 vs. B2.* This display allows users to compare the scenarios developed using the A2 estimates of greenhouse gas emissions versus those developed using the B2 estimates. The user has the option of displaying the A2 versus B2 scenarios for the early-, mid-, or late-century periods.

- *Trend 1990 to 2099.* Users can view the projected change in a climate parameter for 20-year overlapping periods beginning from 1990 to 2009 and continuing to 2080 to 2099. Uncertainty is communicated by including on the display the maximum and minimum projected change across the entire scenario ensemble. In addition, the 25th and 75th percentiles of the projected changes are shown.

To assist users in interpreting the information provided, each display is accompanied by an audiovisual learning guide composed of a set of topics or modules. An example for the "Mid versus Late Century" display is shown in figure 6. The "Essentials" topic provides general information on the intent of the display, explains the axes of the graph, and provides examples on how to read the graph. The next three topics provide background information on important terms and concepts necessary for interpreting the display—including, for this example, definitions of histograms, ensembles, and average values. The last two topics are designed to help users better understand the rationale behind how the scenarios were constructed and the

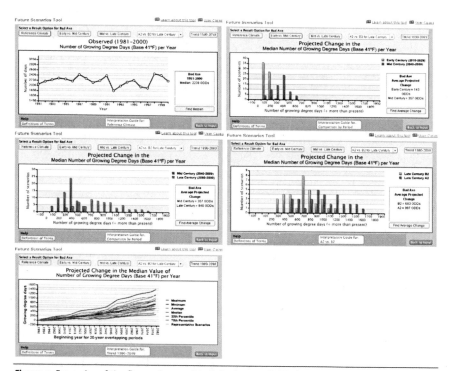

Figure 5. Examples of the five major graphical displays available from the Future Scenarios Tool

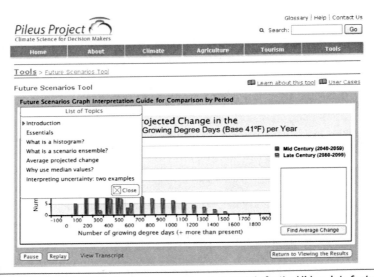

Figure 6. The topics included in the audio–visual learning module for the Mid vs. Late Century display

uncertainty surrounding the scenarios. Similar audiovisual learning guides are provided for the other displays.

An additional aid is a set of "user cases" that were developed to provide examples of how the future scenario ensembles could be used in planning and decision making. The user cases were selected to represent a broad range of possible uses, including uses beyond those related to the subindustries that were the focus of the Pileus Project. The set of user cases includes:

- *Case 1:* How might heat accumulation increase during the growing season, and what are the implications for corn production?
- *Case 2:* How might changes in the length of the growing season affect the feasibility of double cropping?
- *Case 3:* How might the risk of damaging springtime cold temperatures change in the future?
- *Case 4:* How might the date of first fall freeze change in the future, and what are the implications for vinifera grape production in Michigan?
- *Case 5:* What are some of the implications for tourism of changes in the length of wet-day sequences?
- *Case 6:* How might the requirements for air conditioning change in the future, and what are the implications for energy use?

SHELF LIFE OF THE CLIMATE SCENARIOS

One aspect of climate scenarios that is rarely addressed in the literature is their "shelf life." This is an important concern because of the continuous improvement of GCMs and the development of new downscaling approaches and alternative estimates of future greenhouse gas emissions. Users of climate scenarios are faced with the question of whether previously developed scenarios based on earlier GCM simulations, downscaling approaches, or emission estimates are still suitable for a planned assessment, or if, at considerable cost in time and resources, new scenarios should be developed.

To address this question, the scenarios developed for the Pileus Project were compared to a six-member ensemble developed earlier based on simulations from GCMs that were run for the IPCC Second Assessment. The example shown in figure 7 is for the projected change in the date of last spring freeze at Eau Claire, Michigan, for the mid-century period of 2040–2059. The range of uncertainty for the scenarios used in the Pileus Project is shown by the histogram (vertical bars) on the figure, and the range of uncertainty for the scenarios developed from the older GCM simulations is shown by the horizontal bar. There is considerable overlap in the uncertainty ranges for the two scenario suites, although the older scenarios suggest larger projected changes compared to the more recently developed scenarios. Similar patterns were found for other parameters (not shown). These comparisons suggest that a scenario ensemble is more robust than an individual scenario and likely has a greater shelf life, and that it is imperative

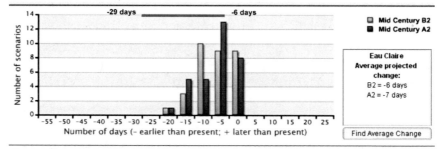

Figure 7. Comparison of future climate scenarios of the projected change in the median date of last spring freeze by 2040–2059 for Eau Claire, Michigan. The projected changes for scenarios developed from GCM simulations used for the IPCC Third Assessment are shown by the histogram. The range of projected changes for scenarios developed from GCM simulations used for the IPCC Second Assessment is shown by the horizontal line.

to include as many GCM simulations, downscaling methodologies, and emission scenarios as possible when developing a scenario ensemble to capture the uncertainty range. Furthermore, older and more recent scenario generations could potentially be combined to better estimate the uncertainty range.

CONCLUDING REMARKS

The intent of this chapter was to describe the myriad considerations and challenges encountered when developing, distributing, and communicating future climate scenarios at the local spatial scale for climate change assessments. Another goal was to illustrate the many advantages of scenario ensembles for assessment studies and decision making. An ensemble allows for a quantifiable range of uncertainty to be estimated, and scenario ensembles have longer "shelf lives" compared to individual scenarios. Furthermore, climate scenarios are much more useful for decision making if they are accompanied by an estimation of the uncertainty in the projected future climate. The design and construction of climate scenarios involves a large amount of resources, making it imperative that scenarios are relevant and accessible to a range of potential users. However, designing a delivery system that facilitates access to a large ensemble of scenarios while at the same time providing sufficient background information for decision makers to correctly interpret and responsibly use the scenarios is a challenging undertaking. The Future Scenarios Tool developed for the Pileus Project is one example of a web-based delivery system. Readers are encouraged to pursue and make use of the scenarios available at www.pileus.msu.edu.

NOTE

The Pileus Project was funded by the U.S. Environmental Protection Agency, project number R83081401–0. This paper has not been subjected to peer review by this agency. The U.S. Environmental Protection Agency does not endorse any of the materials that are being shown or advertised on the Pileus Project website. In addition, the views expressed in this manuscript are those of the authors and do not reflect the views or policies of the U.S. Environmental Protection Agency. The authors are solely responsible for any errors or omissions.

The Pileus Project was the joint effort of many researchers and decision makers. The principal investigators and collaborators on the project were Jeffrey Andresen, J. Roy Black, Donald Holecek, Sarah Nicholls, Peter Sousounis, and Julie Winkler. Jeanne Bisanz was the project administrator and leader of the web-design team; Lori Langone provided administrative assistance for the tourism portion of the project. Post-docs, graduate students, and undergraduate students who contributed to the project include Tracy Beedy, Johnathan Clark, Galina Guentchev, Hai Kyung Min, Jeonghee Noh, Krerk Piromsopa, Haryono Prawiranata, Charles Shih, Ryan Torre, Jenni van Ravensway, and Costanza Zavalloni.

The authors thank Jeff Andresen, Aaron Pollyea, and Peter Kurtz of the Michigan State Climatology Office for providing the historical climate observations for the Michigan stations. We would also like to acknowledge the foresight, guidance, and investment provided by Fred Nurnberger (past state climatologist) and Jeff Andresen (current state climatologist) for the additional homogeneity and quality-control procedures and checks routinely applied to the observations from Michigan's climate reporting stations. Additionally, we thank Pang-Ning Tan from the Department of Computer Science and Engineering for his helpful input on the development of the wet/dry day scenarios. Thanks also to John Furlow and Jordan West, our EPA project managers, for their assistance and guidance. Finally, we thank the many stakeholders who contributed to the Pileus Project.

REFERENCES

Ahmad, Q.K., R.A. Warrick, T.E. Downing et al. Methods and tools. 2001. In *Climate Change 2001: Impacts, Adaptation, and Vulnerability*, ed. by J.J. McCarthy, O.F. Canziani, N.A. Leary, D.J. Dokken, and K.S. White, 105–143. Cambridge: Cambridge University Press.

Antolik, M.S. 2000: An overview of the National Weather Service's centralized statistical quantitative precipitation forecasts. *Journal of Hydrology* 2(39):306–337.

Carter, T., M. Parry, S. Nishioka, and H. Hurasawa. 1996. Technical guidelines for assessing climate change impacts and adaptations. In *Climate Change 1995—Impacts, Adaptations, and Mitigation of Climate Change: Scientific-Technical Analyses*, ed. by R.T. Watson, M.C. Zinyowera, and R.H. Moss, 823–833. Cambridge: Cambridge University Press.

Carter, T.R., R.N. Jones, X. Lu, S. Bhadwal, C. Conde, L.O. Mearns, B.C. O'Neill, MDA, Rounsevell, M.B. Zurek 2007. New assessment methods and the characterisation of future conditions. In *Climate Change 2007: Impacts, Adaptation and Vulnerability*, ed. by M.L. Parry, O.F. Canziani, J.P. Palutikof, P.J. van der Linden, C.E. Hanson, 133–171. Cambridge: Cambridge University Press.

CCSP. 2008. *Climate Models: An Assessment of Strengths and Limitations*. A Report by the U.S. Climate Change Science Program and the Subcommittee on Global Change Research [D.C. Bader, C. Covey, W.J. Gutowski Jr. et al.]. Washington, DC: Department of Energy, Office of Biological and Environmental Research.

Christensen, J.H.B., A. Hewitson, A. Busuioc et al. 2007. Regional climate projections. In *Climate Change 2007: The Physical Science Basis*, ed. by S. Solomon, D. Qin, M. Manning et al. Cambridge: Cambridge University Press.

Intergovermental Panel on Climate Change (IPCC). 2001. *Climate Change 2001: The Scientific Basis*. Cambridge: Cambridge University Press.

Kalnay, E., M. Kanamitsu, R. Kistler et al. 1996. The NCEP/NCAR 40-year reanalysis project. *Bulletin of the American Meteorological Society* 77:437–471.

Jones, R.N. 2000. Managing uncertainty in climate change projections: Issues for impact assessment. *Climatic Change* 45:403–419.

Mearns, L.O., M. Hulme, T.R. Carter, R. Leemans, M. Lal, and P. Whetton. 2001. Climate scenario development. In *Climate Change 2001: The Scientific Basis*, ed. by J.T. Houghton, Y. Ding, D.J. Griggs et al., 739–768. Cambridge and New York: Cambridge University Press.

Nakicenovic, N. et al. 2000. *IPCC Special Report on Emissions Scenarios*. Cambridge: Cambridge University Press.

Palutikof, J.P., J.A. Winkler, C.M. Goodess, and J.A. Andresen. 1997. The simulation of daily temperature time series from GCM output. Part I: Comparison of model data with observations. *Journal of Climate* 10:2497–2513.

Qian, B., S. Gameda, and H. Hayhoe. 2008. Performance of stochastic weather generators LARS-WG and AAFC-WG for reproducing daily extremes of diverse Canadian climates. *Climate Research* 37:17–33.

White, M.A., N.S. Diffenbaugh, G.V. Jones, J.S. Pal, and F. Giorgi. 2006. Extreme heat reduces and shifts United States premium wine production in the 21st century. *Proceedings of the National Academy of Sciences* 103:11217–11222.

Winkler, J.A., J.P. Palutikof, J.A. Andresen, and C.M. Goodess. 1997. The simulation of daily temperature time series from GCM output. Part II: Sensitivity analysis of an empirical transfer function methodology. *Journal of Climate* 10:2514–2532.

Preparing for Climate Change in the Great Lakes Region

THERE IS AMPLE EVIDENCE THAT CLIMATE CHANGE COULD HAVE SERIous repercussions on the economy, ecology, infrastructure, and lifestyles of the Great Lakes region. Sooner or later, residents of the region will have to adapt to current and anticipated conditions caused by changes in temperature, precipitation and other seasonal weather patterns. Adapting to climate change requires people to make decisions under conditions of uncertainty regarding the probability, magnitude, and timing of possible effects. Fortunately, researchers have begun to address how people make decisions in similar circumstances and apply those lessons to the challenges of climate change adaptation.

This volume is part of a broader effort by Michigan State University's Environmental Science and Policy Program (ESPP) to spur information exchange and discussion about these topics among diverse stakeholders in the Great Lakes region. These stakeholders include academic researchers from the social and natural sciences, representatives of state and federal agencies, corporate and small-business interests, farmers and other land managers, and a variety of nongovernmental organizations. This book is comprised of updated papers from a symposium at MSU in March 2007. Follow-up conferences were held in 2008 and 2010. At all of these gatherings, the diverse speakers and attendees participated in discussions of the key messages to be heeded regarding climate change in the Great Lakes region.

During the 2009–2010 academic year, MSU, with support from the Michigan Agricultural Experiment Station, held a series of distinguished lecturers on climate change and engaged a group of Climate Policy Fellows in ongoing discussions with each other and with the visiting lecturers to help identify key concerns. The fellows, emerging leaders from state and local government, the private sector, and nongovernmental organizations, provided further insights into the kinds of information and activities needed to respond to climate change in the region. They also form the core of an ongoing network that will help inform discussions of, and shape responses to, climate change in this region and beyond.

In this final chapter, we have tried to summarize and synthesize the key concepts that emerged from our discussions of the last five years and that guide our ongoing efforts to link climate change research and scholarship on decision making. This chapter goes beyond a simple retelling of the hypotheses and conclusions of the preceding chapters. Rather, it provides broader conclusions about the nature of climate change in the region, the challenges of adapting to these, and suggestions for how to address these challenges. It builds not only on the materials of the preceding chapters, but on the extensive discussion among decision makers and scientists that were centerpieces of these activities.

The key messages are organized into four general categories: (1) the current and anticipated effects of climate change in the Great Lakes region, (2) the relationship between uncertainty and climate-relevant decision making, (3) needs for additional research, and (4) prescriptions for action. We hope these insights can continue to shape a productive dialogue on how stakeholders in the region can address climate change in an effective and efficient manner. We further hope that the concepts listed here can serve as a jumping-off point for discussions in other regions and at other scales of consideration. And of course, we firmly believe that these messages provide the basis for action.

EFFECTS OF CLIMATE CHANGE IN THE GREAT LAKES REGION

There is meteorological and biological evidence that climate change is occurring in the Great Lakes region. This evidence includes changed precipitation patterns, reduced freezing of the Great Lakes, and greater variability in seasonal weather. Moreover, many species have shifted their distribution or timing of their migration. These changes have the potential to create major disruptions in ecological systems, and thus the economic and social systems that rely upon them. The economy of the Great Lakes region and well-being of its residents depend heavily on weather-sensitive sectors of agriculture and natural-resources-based recreation.

Models indicate that even under the most aggressive mitigation scenarios (i.e., we sharply reduce emissions of climate-changing greenhouse gases), our climate will continue to change in the coming decades. This is due, in part, to the concentrations of greenhouse gases that have already been released into the atmosphere. In addition, even if popular and political support for renewable energy sources and efficiency continues to grow, there will be a lengthy transitional period in which industrialized nations will continue to rely on fossil fuels. Moreover, worldwide energy demands continue to grow, meaning that a mix of fuel types likely will be required to meet power needs.

When considering the climatic changes that await the Great Lakes region, it is important to keep the following three factors in mind:

Effects Are Not Certain

Agriculture, tourism, public health, fisheries, and many other aspects of life in the Great Lakes region are affected by patterns in weather and climate. Changes in climate may have wide-ranging consequences on physical, biological, social, and economic systems. However, the exact effects of climate change are uncertain. Our ability to predict the effects of climate change decreases as we consider longer time frames or smaller geographic scopes. This means that it is difficult to pinpoint how specific locations will be affected in the future. Moreover, because these systems are complex, some of their responses to climate change will be surprising and could be very rapid. The effects are also unlikely to occur on a smooth trajectory; because systems are interrelated, changes in one system could spur a domino effect of changes in other systems.

There Will Be Direct and Indirect Effects

Some effects will stem directly from changes in the climate and weather patterns in the Great Lakes region, including temperature and precipitation. These changes, however, will also spur wide-ranging indirect effects. For example, an increase in the number of extreme heat days could boost electricity demands during summertime months as people rely more heavily on air conditioning. In turn, utilities would have to construct additional power-generation facilities. Warmer wintertime temperatures reduce surface freezing of the Great Lakes, which increases annual evaporation and may change lake levels. Changes would have indirect effects on fish populations if nearshore spawning areas are affected.

Still other effects could result from pressures exerted by climate-related changes felt outside the Great Lakes region. As increased summer heat and drought hit other parts of the United States or abroad, there could be greater demand for Great Lakes water resources, or increased development pressure as people migrate to cooler regions in the north. Although some indirect effects can be predicted and modeled, we are likely to be caught off guard by others.

Effects Will Not Be Experienced Equally

There will be both costs and benefits associated with climate change in the Great Lakes region, but these effects will not be distributed evenly. Due to the complex interactions of geography and climate, some local areas will experience few changes. On the other hand, other areas might experience numerous and dramatic differences.

Moreover, the seriousness of climate change depends in part on the nature of who or what is affected. Some systems, species, and individuals are more vulnerable to changes in the climate. Others are more resilient. A slight change in water temperature might doom one species of fish, while another is able to flourish in a wide range of habitat.

One key area of difference is in the availability of monetary and other resources among different municipalities, social subgroups, and individuals. Some forms of adaptation require significant investments. Entities with access to substantial resources are better positioned to make the changes necessary to avoid significant disruptions or heartache. Those without resources will experience greater suffering.

UNCERTAINTY AND CLIMATE-RELATED DECISIONS

One topic often revisited in discussions at MSU events is how people confront the uncertainties inherent in climate change. An overriding theme is that uncertainty is not unique to climate change.

Stakeholders Are Familiar with Uncertainty

Decision makers in the Great Lakes region already confront substantial uncertainty, which comes from a variety of sources. Some of this uncertainly is related to the climate. Climate-vulnerable systems already face difficult-to-predict variability in the weather. For example, farmers must steel themselves for unforeseen drought and floods, and city managers must budget for annual snow removal long before the first flakes fall.

Uncertainty comes from a variety of forces. Political processes, demographic trends, fads and fashion, and technological advances all influence economic and social-systems change in unforeseen ways. All of these factors complicate decision making and may determine the ultimate success of any given strategy. Some sectors have developed successful strategies to cope with uncertainty. For example, farmers rely heavily on crop insurance to mitigate losses stemming from severe weather.

Uncertainty Is Interactive

The uncertainties in one system may make it more difficult to predict changes in another system. The resilience or vulnerability of a particular social or ecological system to climate change depends, in part, on changes in non-climate-related aspects of the system. For example, the vulnerability of farmers depends on the types of crops they grow, and their choice of crops is influenced by markets, which are affected by food trends. At times, it is difficult for a stakeholder to predict or assess the seriousness of potential climate change, due to the uncertainty in related systems that have nothing to do with climate.

Uncertainty Is Just One Obstacle to Adaptation

There are many potential constraints to planning and implementing responses to climate change. These obstacles include a lack of resources to make necessary changes, poor institutional support, or the need to attend to other priorities. While it is important to understand how uncertainty influences decision making, it must be recognized as just one of many factors that limit the ability of stakeholders to adapt successfully.

RESEARCH NEEDS

One of the questions driving our climate-related events with stakeholders has been how academic and government research can better satisfy stakeholders' decision-making needs.

Technical Information Isn't Everything

A wide variety of research is needed to better understand potential effects of a shifting climate and the best ways to address these changes. Stakeholders are interested in scientific analyses that help them predict local and regional climate change and how those changes will affect Great Lakes species and ecosystems. While scientific information is critical to decision processes, however, climate-related risk management will be guided by human values and often rely on relatively simple heuristics rather than elaborate calculations. Research is needed to better understand the variable ways in which decision makers currently make climate-related decisions and respond to changing environmental conditions.

Interdisciplinary Scholarship Is Needed

Because climate, ecological, and social systems are complex, the most successful climate-related research projects will integrate knowledge and expertise from across disciplines. The integration of different perspectives will also foster more creative solutions to the challenges of climate change.

Stakeholders Should Be Involved in Research

Scientific models and data can be useful in decision making if they are presented in a manner that takes account of the objectives, interests, and communication needs of decision makers. One challenge is to effectively communicate high-consequence risks that have a low probability of occurring. Ongoing, trustful relationships between researchers and decision makers will facilitate the creation of useful climate-related information. Naturally, scientific expertise is critical to understanding climate change and developing solutions. However, traditional forms of knowledge are also vital to fully understand how ecological and social systems are affected and can respond to changes. Sources of traditional knowledge include Native American tribes, farmers, and other stewards of land and water in the region.

PRESCRIPTIONS FOR ACTION

Continued research and analysis is critical, but uncertainties should not be a barrier to action.

Think and Act Locally and Regionally

While instituting responses to climate change may be more effective or feasible at a local level, these efforts must be coordinated in order to avoid redundancies or contradictory actions. Planning for climate change should be conducted through nested processes, in which local actions are coordinated at a regional scale.

Assessment Matters

In order to determine vulnerabilities and set priorities for adaptation, place-based assessments of climate vulnerability should be conducted at multiple scales. These assessments would identify which sectors are most vulnerable to climate variability and how decision makers can get the "greatest bang for their buck." Participatory stakeholder-researcher partnerships can contribute to successful assessments. Stakeholders can help to identify vulnerabilities and establish priorities for action.

Focus on Resilience

Under conditions of uncertainty, increasing the resilience of particularly vulnerable systems may provide some protection against serious climate-related problems. This might mean addressing non-climate-related factors that are acting as stressors to critical systems. For example, populations of threatened species could be bolstered to better withstand potential climate-related crises. Also, local economies that rely heavily on a climate-vulnerable industry could stress diversification in economic development plans.

Climate Change Must Be Integrated into Existing Decision Processes

Climate change cannot be addressed in isolation. Climate-related adaptation is unlikely to be successful if treated as a separate institutional program or planning process. Rather, the potential effects of climate change must be considered as part of other long-range planning processes. This is because climate change is just one of many interrelated stressors facing natural and human systems. One cannot plan for the conservation of a particular species without considering how their long-term survival will be affected by changing distribution of habitat. Stormwater management efforts must consider potential changes in precipitation patterns, and infrastructure must be built to accommodate a wide range of climate conditions.

Adaptation Is Not a One-Time Decision

Due to the uncertainty inherent in climate systems, stakeholders should use iterative decision-making models so they do not become locked into maladaptations. Due to uncertainties inherent to climate science and the interactive nature of climate-sensitive systems, flexibility should be a key component of addressing climate change. Successful approaches will include continued monitoring of changes, evaluation of adaptive responses, and revisiting critical decisions on a regular basis. In this way, addressing climate change should be seen as adaptive risk management.

MOVING FORWARD

These principles are grounded in the experience of researchers and stakeholders in the Great Lakes region. But comparison with the messages from recent national assessments of how to respond to climate change suggests that our experiences are not unique (U.S. National Research Council 2010a, 2010b) and that our region can learn from and help inform the rest of the nation. Decision makers in federal, state, tribal, and local governments, in small businesses and large corporations, as well as individuals and households, will be making myriad decisions that will affect and be affected by climate change. The challenge, embodied in the above principles, is to effectively link science and decision making as we respond to climate change (see also Mastrandrea, Heller, Root, and Schneider 2010; Smit and Wandel 2006; Vogel, Moser, Kasperson, and Debelko 2007).

In response to this challenge, Michigan State University and the University of Michigan have partnered to form the Great Lakes Integrated Science and Assessments Center (GLISA). GLISA is supported by the National Oceanographic and Atmospheric Administration and is one of a network of 11 Regional Integrated Science and Assessment Centers intended to help usher in the new era of climate change research. This new institution is a centerpiece of those efforts and in that sense is a continuation of the work reported throughout this volume. In fact, several of our contributors are active leaders in GLISA (further information is available at glisa.umich.edu and glisa.msu.edu). GLISA serves as a node in a network of researchers and decision makers that span the Great Lakes region, and strives to strengthen that network. Guided by stakeholder input, GLISA funds research focused towards addressing uncertainties and aiding climate-sensitive decisions. In that sense, it reflects the core message of this volume—that we have entered a new era in responding to climate change, an era that requires ongoing interaction between science and decision making.

REFERENCES

Mastrandrea, M.D., N.E. Heller, T.L. Root, and S.H. Schneider. 2010. Bridging the gap: Linking climate-impacts research with adaptation planning and management. *Climatic Change* 100:87–101.

Smit, B., and J. Wandel. 2006. Adaptation, adaptive capacity and vulnerability. *Global Environmental Change* 16:282–292.

U.S. National Research Council. 2010a. *Adapting to the Impacts of Climate Change.* Washington, DC: National Academies Press.

U.S. National Research Council. 2010b. *Informing an Effective Response to Climate Change.* Washington, DC: National Academies Press.

Vogel, C., S.C. Moser, R.E. Kasperson, and G.D. Dabelko. 2007. Linking vulnerability, adaptation, and resilience science to practice: Pathways, players, and partnerships. *Global Environmental Change* 17:349–364.

Notes on the Contributors

Jeffrey A. Andresen is associate professor of meteorology/climatology with Michigan State University's Department of Geography and the state climatologist for Michigan. He currently serves as director of the Michigan Climatological Resources Program and co-director of the state's Enviroweather system, which supports agricultural pest, production, and natural resource management decision making across Michigan. The primary focus of Andresen's research has been the influence of weather and climate on agriculture, especially within Michigan and the Great Lakes Region. He holds an undergraduate degree from Northern Illinois University in the field of meteorology and M.S. and Ph.D. degrees from Purdue University in the field of agricultural meteorology/climatology.

Suzanne Belliveau completed a Master's degree in geography at the University of Guelph focused on climate change adaptation in agriculture, and she subsequently worked as a research associate in the department. She now works in the field of disaster preparedness and response.

David Bidwell currently serves as the program manager for the Great Lakes Integrated Sciences and Assessments Center. His research interests revolve around environmental decision making. Bidwell holds an M.S. in natural resource policy from the University of Michigan and a Ph.D. in sociology from Michigan State University.

Jeanne M. Bisanz served as the Project Manager for the Pileus Project and coordinated the development of web-based tools for stakeholders. Previously, she was the regional coordinator for the Great Lakes Regional Assessment, which was conducted as part of the United States National Climate Change Assessment.

Ben Bradshaw is an associate professor of geography at the University of Guelph. His research seeks to identify the economic, political, and cultural determinants of environmental degradation in Western society, and the various tools of governance that might best alleviate such degradation. His doctoral work on the environmental implications of subsidy reform in the agriculture sector led to further agriculture-centered research, featured herein, on the implications of climate change for farm-level decision making.

Johnathan Clark is a graduate student studying atmospheric physics at Howard University. Clark received his bachelor of science in earth science from Michigan State University. He currently is involved with the modeling of mesoscale convective systems in West Africa.

Thomas Dietz is a professor of sociology and environmental science and policy, serves as assistant vice president for environmental research at Michigan State University, and co-directs the Great Lakes Integrated Sciences and Assessments Center. He holds a Ph.D. in ecology from the University of California, Davis, and a bachelor of general studies from Kent State University. Dietz has served as the chair of the U.S. National Research Council Committee on Human Dimensions of Global Change and as vice chair of the Panel on Advancing the Science of Climate Change of America's Climate Choices study.

William E. Easterling is professor of geography and dean of the College of Earth and Mineral Sciences at Penn State University. His research focuses on the consequences of climate change for food security in different regions of the world. He was one of two coordinating lead authors on the agriculture and forestry chapter in the Fourth Assessment Report of the Intergovernmental Panel on Climate Change.

Galina S. Guentchev is a UCAR VSP Visiting Scholar. She obtained her M.Sc. degree at Sofia University, Bulgaria, and her Ph.D. at Michigan State University. Her research interests are related to the impacts of regional climate variability and change on human and natural systems.

Kimberly R. Hall is the Great Lakes climate change ecologist for The Nature Conservancy, based in Lansing, Michigan. She holds an M.S. and a Ph.D. from the University of Michigan, where she conducted research on forest songbirds. Her work at The Nature Conservancy focuses on synthesizing information on climate change impacts and vulnerabilities, and on working collaboratively with colleagues and partners on adaptation strategies that benefit biodiversity and people in the Great Lakes region.

Scudder D. Mackey is principal and owner of Habitat Solutions NA, an environmental consulting firm based in Chicago, Illinois. Mackey holds a doctorate in geology with areas of technical specialization in aquatic habitat characterization and mapping; development of biophysical linkages to habitat; and climate change impacts to nearshore, coastal, and riverine processes and habitats. Recent work has focused on the potential effects of altered water level and flow regimes on aquatic habitat distribution, pattern, and connectivity.

Sabine M. Marx is an associate research scientist and the managing director at the Center for Research on Environmental Decisions at Columbia University. Her work falls in the area of decision making under uncertainty, with a particular focus on the use of climate information in agriculture, public health, and disaster preparedness and management.

Hai Kyung Min is a web architect at Google. She received her Bachelor of Art degree in mass communication from Ewha Woman's University (Seoul, Korea) and her M.A. in telecommunication, information studies, and media (concentration on digital media art and technology) from Michigan State University.

Susanne Moser is director of Susanne Moser Research & Consulting, a social science research fellow at Stanford University's Woods Institute for the Environment, and a research associate at the University of California-Santa Cruz. Her work focuses on adaptation to climate change in the coastal, health, and forest sectors; resilience in the face of hazards and global change; decision support; and effective climate change communication in support of social change.

Xianzeng Niu is a soil scientist in the Earth and Environmental Systems Institute at The Pennsylvania State University. Since joining Penn State in 2005, he has participated in a variety of research projects related to climate change impact, mitigation, and adaptation studies, including applying remote sensing, GIS, and modeling techniques to reliability and uncertainty analysis of crop models in simulation of large-area crop yields, bioenergy production, land-use change, and their impacts on terrestrial carbon sequestration.

Krerk Piromsopa is an assistant professor of computer engineering at Chulalongkorn University, Thailand. He received his bachelor of engineering and master of engineering in computer engineering, both from Chulalongkorn University, and his Ph.D. in computer science from Michigan State University. Piromsopa's research focuses on computer architecture, computer security, digital synthesis, and high performance computing. He is also interested in climate scenario development.

Haryono Prawiranata is an environmental engineer with professional experience in numerical modeling. He currently serves as a transportation and air quality modeler for the Tri-County Regional Planning Commission in Lansing, Michigan.

Jenni van Ravensway received her B.S. and M.S. degrees in geography from Michigan State University. She is currently a research associate at the Center for Global Change and Earth Observations at Michigan State University. Her research interests involve the development and application of geospatial technologies to analyze global change issues that integrate social, terrestrial, and climate systems, with a focus in remote sensing and GIS.

Terry L. Root is well known for her capacity to reach decision makers and the general public, having published extensively and received several awards, including being part of the Intergovernmental Panel for Climate Change that shared the 2007 Nobel Peace Prize with former Vice President Al Gore. She is a professor, by courtesy, of biology and senior fellow at the Woods Institute for the Environment at Stanford University, and a fellow at the California Academy of Sciences. She earned her undergraduate degree from the University of New Mexico, her master's degree from the University of Colorado, and her doctorate from Princeton University.

Joel D. Scheraga is the Senior Advisor for Climate Adaptation in the U.S. Environmental Protection Agency's Office of Policy in the Office of the Administrator. His responsibilities include helping the EPA design and implement effective climate change adaptation measures to protect human health and the environment. Scheraga served as a lead author and contributing author for the Intergovernmental Panel on Climate Change, which was awarded the 2007 Nobel Peace Prize.

Clark Seipt is program coordinator for the International START (Global Change SySTem for Analysis, Research, and Training) Secretariat. She also serves as staff coordinator for day-to-day activities. Seipt received her M.S. in geography from The Pennsylvania State University with a thesis on the utility of seasonal climate forecasts for Argentinean agriculture. She has a B.S. in Environmental Sciences and Spanish from the University of Virginia.

Barry Smit is a professor of geography and Canada Research Chair in Global Environmental Change at the University of Guelph. His research explores the vulnerability and adaptations of communities and socio-economic systems to global environmental changes. This interdisciplinary and applied research has been undertaken in numerous developed and developing countries and has contributed to notable publications including the 2001 and 2007 IPCC Assessment Reports and Ontario's Expert Panel Report on Climate Change Adaptation.

Adam Terando is a climate scientist and research coordinator in the Department of Biology at North Carolina State University. His research focuses on the impacts of climate change on ecosystems and agro-ecosystems and the complex human-environment relationships that drive these processes. Recent work includes evaluation of global climate model performance in the context of regional agro-climate indices, Bayesian data-model fusion to appropriately weight and downscale GCMs and account for model uncertainty, and modeling the effect of path dependence on farmers' ability to adapt to climate change.

Ryan S. Torre received his B.A. and M.A. in telecommunications, both from Michigan State University. His responsibilities for the Pileus Project included interface design and programming for several of the Flash-based tools, including the Future Climate Scenarios Tool.

Elke U. Weber is the Jerome A. Chazen professor of international business and professor of psychology at Columbia University. She is an expert on behavioral models of judgment and decision making under risk and uncertainty, specifically in financial and environmental contexts. At Columbia, she founded and co-directs the Center for Research on Environmental Decisions, which investigates ways of facilitating human adaptation to climate change and climate variability. She has served on several National Academy of Sciences advisory committees related to human dimensions of global change, was a member on an American Psychological Association Task Force that issued a report on the Interface between Psychology and Global Climate Change, and is a lead author in Working Group III for the 5th Assessment Report of the Intergovernmental Panel on Climate Change.

Julie A. Winkler is a professor of geography at Michigan State University. She received her bachelor of science degree in geography from the University of North Dakota, and her M.A. and Ph.D. in geography, both from the University of Minnesota. Winkler's research focuses on several aspects of geography and climatology including synoptic and applied climatology, regional climate change, and climate scenario development.

Reflections on
Stephen H. Schneider

STEPHEN H. SCHNEIDER WAS A LEADING PARTICIPANT IN THE SYMPO-
sium generating the papers in this volume. Steve contributed deep insights,
an unparalleled ability to span disciplines, a measured sense of urgency, an
insistence that we heed the science and be open and clear about uncertainty,
and an unfailing sense of humor in the face of the often daunting climate
problem, his own health struggles, and the increasingly nasty attacks on
climate science. We had hoped that Steve could write our preface, but his
untimely death in July of 2010 intervened before that could happen.

The volume reflects Steve's intellectual breadth—his contributions to
climate research spanned a vast range from physical climatology, to climate
decision analysis, to the sociology of climate science. With Richard Moss,
he created the norm that statements about climate change should always
include assessments of the level of confidence scientists have in their conclusions. The emphasis in this volume on addressing uncertainty reflects his
insights on this issue. His vast intellectual scope was reflected not only in his
published research and in his professional leadership but also in his role as
a public intellectual always teaching those who heard him speak. His journal, *Climatic Change*, was a pioneer in welcoming the full suite of climate
change disciplines, including the physical, ecological, and social sciences.
He was unflagging in his advocacy for interdisciplinarity—across both the
physical and social sciences—in the study of climate change. It is no surprise
to those who had the privilege of knowing him that his last trip was to a
meeting of the International Sociological Association—a venue few other
climate scientists have visited. Not only did he give a keynote address there,
but he also engaged in and stimulated a wide range of discussions among
the sociologists present.

We miss Steve as a path-breaking scientist, as one of our most articu-
late interpreters of the implications of our complex science in both popular

writing and public appearances, and as a warm and humorous presence at our gatherings. Though a remarkably accomplished scientist, Steve was hardly unidimensional—some recall his cameo role in the Woody Allen movie *Sleeper*—and Dietz fondly remembers his last conversation with Steve, which ranged from the underpinnings of climate denial to the nuances of describing uncertainty, but centered on the merits of Russian River Valley Pinots and Chardonnays.

Index

adaptation, 2–3, 66, 129, 146, 249, 256
adaptive capacity, 7–8, 145, 159–60, 163, 173–74, 222
adaptive management, 57, 143, 222
agricultural impacts, 30–31, 135–50, 159–74
agricultural yield, 191, 171–72, 193
aquatic ecosystems, 222–23; habitats, 39, 46, 51–53, 56, 58, 87; species, 46, 66, 70–71, 73. *See also* fish
assessments, climate, 231–32, 234, 255. *See also* Great Lakes Regional Assessment and U.S. National Assessment

behavior, human, 105, 160
best management practices, 84
biofuels, 73, 134–35, 160
bioterrorism, 137
birds, 68, 70, 73, 75, 78–80, 89, 191; migration, 79–80

cascading impacts, 35, 46, 51, 80, 214
choice theory, 106, 114, 116
climate smart, 85, 87, 89
CO2 effects, 31, 67, 85, 89, 134, 139–40; climate-sensitive optimization, 190; multiattribute utility theory, 115; phenology, 76
coastal areas, 41, 84
contamination, 51. *See also* pollution
cost effectiveness. *See* economic impacts and factors
costs. *See* economic impacts and factors
CSS (combined sewer systems) and CSO (Combined Sewer Overflow), 119, 217–20, 226–27, 229

decision support, 57, 120, 183, 202–3, 213–14, 221–26, 234
decisionmakers. *See* stakeholders
decisionmaking, 7, 99–123, 151, 179–205, 213–26; integration with science, 179–205, 221; process, 181–86, 203, 213, 223, 256; tools, 222–23; tradeoffs in, 115–17, 152, 191
decisions, 89, 213–27, 239–40, 249–57; adaptation, 129–53, 159–74; evaluation, 190, 196; goal-oriented versus evaluative, 7, 194; iterative, 183, 201, 206, 256; one-time versus non-final, 7, 194, 196; optimization, 189–90, 196; protection of ecosystem and species 83–84, 87; short-term versus long-term, 7, 52, 115, 192–94, 196; time horizon of, 121, 186, 188, 192–94, 196, 199; uncertainty, 99–123, 179–205
discounting, 112–14, 118, 120–21
downscaling, 8, 10, 43, 121–22, 201, 214, 232, 236–40, 245–46
drought, 82, 119, 222, 252–53, biodiversity, 69, 73–74, 164–74; extreme event, 84, 108, 137, 148
DUST (Decision Uncertainty Screening Tool), 7–8, 179–205

ecological impacts, 35, 51, 54
economic impacts and factors, 116–19, 135, 138, 184–85, 188; risks, 122, 161, 165, 168–69
ecosystem shift. *See* species change
energy, 35–58, 63–89, 134, 137–38, 233, 251. *See also* biofuels
environmental zones, 39

feedback loops, 65
fish, 4, 78, 184–85; cold-water, 65, 68, 70; Great Lakes, 38–39, 48–49, 51–52; 56, 85; 251–52; Lake St. Clair, 46, 49–50
food security, 6, 119, 130, 134, 159
framing, 111, 117, 122, 203

GCM (general circulation model/global climate model), 67, 76, 83, 134, 149, 187, 214–16, 219–220, 232, 236–40, 245–46
geography, 3, 73, 104, 235, 252
geology, 35–38
GIS (Geographic Information System), 105
GLISA (Great Lakes Integrated Sciences and Assessments Center), 11, 257
Great Lakes Regional Assessment, 225, 234
groundwater, 36, 40–41, 73, 129

heuristics, 6, 99, 106, 117, 254
hydrology, 35–36, 38, 40–41, 65, 220

ice cover, 22, 24, 28, 40, 65, 82–83
impact assessment, 131–32, 150, 159–74
incentives. *See* economic impacts and factors
insects, 5, 68, 70, 79–81, 137, 165
integrated assessment, 139–40, 202. *See also* assessments, climate
IPCC (Intergovernmental Panel on Climate Change), 64, 104, 134, 138, 153, 239, 245
irrigation, 31, 119–20, 129, 138, 142, 169, 191, 193

Kyoto Treaty, 116

lake effect, 4, 18–20, 26, 28–29, 32
Lake Erie, 42–43, 48–49, 58, 66, 79, 85, 162; "dead zone," 49, 66
Lake Huron, 42–43, 64, 71
lake levels, 40–58, 83, 85, 105, 184–85, 191, 193, 251
Lake Michigan, 43, 64, 71, 88, 119
Lake St. Clair, 42–49, 54, 105
Lake Superior, 64, 66, 193
livestock, 119, 133, 139–42, 160, 167, 169–70

localization, 122. *See also* downscaling

mammals, 69, 71–73
master variables, 35–37, 40–41; natural variables, 36
mitigation, 4, 100, 117, 120, 130, 132, 179–80, 187, 250
modeling, 72, 83, 183, 197, 202, 235, 239; agricultural adaptation, 129–53; climate, 6, 40–42, 204, 214–15, 223; normative versus descriptive, 103–6. *See also* impact assessment

National Hurricane Center, 102

overconfidence, 106–7

paleoclimatological evidence, 20
path dependence, 130, 144–46
pesticides, 133, 137, 139
pests, 63, 74, 137, 139–40, 165–66, 172–73. *See also* insects
phenology, 5, 68, 75–80, 82, 84–87
Pileus Project, 8, 10, 105, 121, 231–47
policy, 100, 115, 120, 129–33, 144; environmental, 118, 121; government, 171, 174; decision making, 179–82, 191, 195, 198–99, 202–4, 213–15
pollution, 36, 65, 84, 113, 119, 223. *See also* contamination
practical application, 8
precipitation, 30, 40–41, 84, 217–20, 231–47; annual, 19, 24–25, 223; effects of increase, 26, 165–66, 173, 222; lake effect, 20, 28–29; patterns, 64, 67, 82, 105, 249–51, 256; timing of, 83, 164, 172
predictability, 6, 99, 100
probability, 101–12, 197, 249, 254
processing: analytical, 102; experiential/affective, 102
public health, 71, 74, 88, 119–20, 141, 194, 216, 217–20, 251

rain. *See* precipitation
resilience, 5, 172, 223, 253, 255
risk, 103; management, 122, 163; perceived, 109–10, 121

scales: spatial, 5, 7–8, 129, 147, 187, 235, 255; time, 20, 129, 184–85
scenario: climate, 145, 160, 231–47; technique, 104
seasonality, 52, 193
sensitivity analysis, 141, 197, 222–23
sewer, 119, 217–20, 226–27, 229
snow cover, 19, 29
snowfall. *See* precipitation
soils, 36, 74, 149, 166
species: invasive, 5, 36, 54, 73, 87–88, 137
species change, 40, 63–65, 67–70, 73–88; evolution, 69–70, 81–82; interaction, 78; monitoring, 77; range and abundance, 70–74, 89; vulnerability, 82, 85
stakeholders, 7–10, 103, 118, 130, 132, 150–53, 183, 185, 194, 198, 214–16, 221, 225–26, 233–34, 241, 249–56
stratification, 65–66, 75

temperature, 21–24, 27–33, 63, 231–48; effects, 67–68, 131, 133–34, 148; extreme, 86–87, 172, 251; water, 64–65
tillage, 119, 129, 138, 142, 164, 168–69, 173
tourism industry, 8, 234–36
trees, 74, 88, 197

uncertainty, 10, 131–32, 214–15, 231–32, 241–46, 249–57; communication in light of, 196–98; in decision-making,

99–123, 179–83, 198, 226, 250; DUST, 7–8, 179–205; fundamental, 6, 130, 135–38, 239; of models, 197; parametric, 6, 130, 135–36, 147–50; structural, 6, 130, 135, 138–47, 239; well-calibrated versus judged versus full, 201, 232
U.S. National Assessment, 225, 234
usable knowledge, 6, 103, 129–53, 215. *See also* utility
USGCRP (United States Global Change Research Program), 7, 219, 221, 223
utility (usefulness), 111–17, 151–52, 215

variability, climate, 104–5, 120, 122, 140, 144–48, 161, 172, 180, 184, 187, 221, 232, 234, 236, 255
variables, 10, 30, 35–38; anthropogenic, 36–37, 40–41, 56–57, 67, 76, 232. *See also* master variables
vulnerability framework, 163

water levels. *See* lake levels
water resources, 35, 41, 52, 83, 135, 222, 252
weather, 10, 26, 137, 171, 180, 235; extreme, 84, 87–88, 110, 159, 192, 223; patterns, 4–5, 18–20; seasonal, 41, 77, 192, 249–51. *See also* drought, temperature
wetlands, 48–50, 52, 73, 84